Atomic Light
(Shadow Optics)

Atomic Light
(Shadow Optics)

Akira Mizuta Lippit

University of Minnesota Press • Minneapolis • London

Published by the University of Minnesota Press
111 Third Avenue South, Suite 290
Minneapolis, MN 55401-2520
http://www.upress.umn.edu

Library of Congress Cataloging-in-Publication Data

Lippit, Akira Mizuta.
    Atomic light (shadow optics) / Akira Mizuta Lippit.
        p.   cm.
    Includes bibliographical references and index.
    ISBN 0-8166-4610-4 (hc : alk. paper) — ISBN 0-8166-4611-2 (pb : alk. paper)
    1. Vision.   I. Title.
    B846.L57   2005
    121'.35—dc22   2005020224

ατομος, *indivisible*

Contents

Acknowledgments

I am deeply grateful to Lisa Cartwright, Laura U. Marks, and Vivian Sobchack, who read this book carefully and offered detailed criticisms. Each made invaluable suggestions that vastly improved the book. Miya Lippit and Albert Liu read the book line by line—they are its coauthors.

I would like to thank Dudley Andrew, Cade Bursell, Michael Fried, Anne Friedberg, Hosokawa Shuhei, Peggy Kamuf, Karatani Kojin, Jean-Claude Lebensztejn, Richard Macksey, Fred Moten, Bill Nichols, Michael Renov, Trinh T. Minh-ha, Hervé Vanel, Catherine Waldby, and Linda Williams, who read the book or parts of it at different stages and offered criticism, references, and lines of inquiry. Catherine Waldby suggested the phrase "catastrophic light," which I have used here. Each has left a singular *exscription* on this book.

Jacques Derrida supported me from an early point in my development, and I am grateful for the kindness he extended to me. Those of us who had the good fortune to work with him will miss his brilliance, generosity, and humor.

In Japan, Iwamoto Kenji and Ukai Satoshi were especially helpful in supervising and critiquing sections of this work. Donald Richie suggested novel directions for me to pursue. Research in Japan was made possible by grants from the Japan Foundation, the Northeast Asia Council of the Association for Asian Studies, San Francisco State University, and the University of California, Irvine, Humanities Center. I am grateful for their support.

Various portions of this book were delivered as lectures at the Center for Japanese Studies at the University of Michigan; the Getty Center; Hitot-subashi University; Indiana University; the Pacific Film Archive; the San Francisco Art Institute; the UCLA Center for Japanese Studies; the University of California, Riverside; the University of Memphis; the University of Pittsburgh; the University of Southern California; Wayne State University; Waseda University; and Yale University. I thank my hosts and interlocutors for their insights and criticisms.

I wish to thank my colleagues and graduate students in the Department of Cinema at San Francisco State University and the Visual Studies Program at the University of California, Irvine, for providing a supportive and stimulating environment, and my colleagues at the UCI Langson Library, especially Dianna Sahhar, for their invaluable research assistance.

The staff at the University of Minnesota Press has been exemplary. Doug Armato edited the manuscript and showed enthusiasm for this project from the beginning; Gretchen Asmussen helped at each stage; Paula Dragosh copyedited the book and strengthened every aspect of it; Sallie Steele indexed the book with precision. I am grateful to each.

0. Universes

The man painted with Chinese ideograms is Hôichi, a blind monk and *biwa* lute player in Kobayashi Masaki's *Kwaidan* (1964, based on Lafcadio Hearn's 1904 collection of "stories and studies of strange things"). The inscriptions, which have been written over his entire body, are Buddhist prayers. In a few moments a phantom will come for Hôichi and escort him to a grave site where he has, for the past several nights, performed an epic poem commemorating the ancient Heike (Taira) clan. Hôichi's song chronicles the 1185 Battle of Danno-ura, where the Heike, including the child emperor Antoku, were annihilated at the hands of their enemy, the Genji (Minamoto) clan. This war marked the end of the classical Heian era and the beginning of the Kamakura shogunate. Hôichi's hosts and audience are the ghosts of Heike, the dead warriors, courtiers, and children who perished in the battle. Unaware of who they are, Hôichi moves closer to them with each performance. He will expire with the song and become one of them upon the completion of his performance. He will cross over to the other side of life and history, into the phantom world of total destruction.

"Hôichi the Earless," in Kobayashi Masaki, *Kwaidan* (*Kaidan*, 1964).

During his extended performance Hôichi has weakened, and his body and complexion bear the signs of illness. His face exhibits an inner anxiety, a shadow darkens his expression. Hôichi's inner dimension is being drawn outward into the absolute exteriority of the dead, of shadows. The priests of Hôichi's temple have noticed his receding energy and realize that Hôichi is being absorbed by a powerful exterior force, by the powerful force of exteriority as such. Unless they intervene, Hôichi will disappear into the outside, his body and vitality swallowed by the phantom world and its destructive energies. To protect Hôichi from the shades and make him invisible to them, the Buddhist priests have covered the outer surface of Hôichi's entire body with prayers. The wet, liquid ink—black except for the red Sanskrit signs on Hôichi's forehead, back, and hands—will plunge the blind monk into a phantom darkness. The ghosts will no longer see Hôichi, just as he does not see them, just as he does not see the world. With these prayers, written on the surface of his body, the entire world will be rendered blind, spectral, invisible.

Hôichi's body is protected by the religious power of the prayers, but also by the materiality of the text. The black ink is a stain that covers and delineates his body like a shadow applied directly onto his skin. It is at once physical and metaphysical: Hôichi's skin the surface where the two dimensions converge. Like an invisible man, Hôichi is made visible in one register by the writing on his body, and invisible in another. The *exscriptions* on Hôichi's body establish two distinct orders of visibility—one phenomenal, the other phantom—and locate him in the interstice between the two. Hôichi is visible in one world, invisible in another. His visuality is doubled, paradoxical, and inside out.

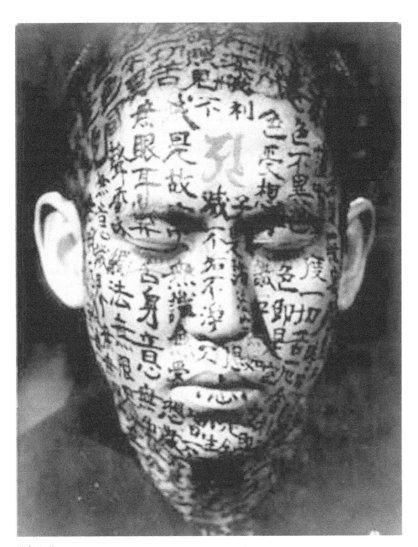

"Hôichi."

The divine script seals Hôichi, his body, and its surface, his skin. With the prayers inscribed or *exscribed* onto his body like an archive, he is hidden within the text, within the archive that covers the surface of his body. It is Hôichi's *body* that must be defended, his material form. By covering Hôichi with prayers, the priests have withdrawn him from the outside world. They have secluded him, hidden him within the archive that, painted onto his skin, is exposed, transposed to the outside. The archive has been *exscribed* onto Hôichi, exposed on his body, which is shielded by and within it. Hôichi has been interiorized by the *exscription*. He is inside out, suspended between the worlds of visibility and invisibility.

The scene of Hôichi's suspended visuality is critical to understanding a crisis initiated at the end of World War II, a crisis in the constitution of the human body. The atomic radiation that ended the war in Japan unleashed an excess visuality that threatened the material and conceptual dimensions of human interiority and exteriority. It assailed the bodies it touched, seared and penetrated them, annihilating the limits that established human existence in the world. The destruction of Hiroshima and Nagasaki in 1945 exposed the fragility of the human surface, the capacity of catastrophic light and lethal radiation to penetrate the human figure at its limit. Under the glare of atomic radiation, the human body was exposed: revealed and opened, but also displaced, thrust outward into the distant reaches of the visible world. It situated the body between not only two worlds but two universes: two separate orders of all things, or even of the same things. Visibility and invisibility, exteriority and interiority, the living and the dead, this world and that other world rest on the surface of Hôichi's body, at the point of contact between the text that covers Hôichi and his skin.

Hovering between visibility and invisibility, outside and inside, life and death, Hôichi's liminal moment is emblematic of postwar Japanese cinema and visuality. Since 1945, the destruction of visual order by the atomic light and force has haunted Japanese visual culture. Yet the origins of this postwar crisis, brought about by the advent of a penetrating radiation,

The atomic bombing of Hiroshima. *Enola Gay* pilot Paul W. Tibbets inscribed the image.

began fifty years earlier in 1895 with the emergence of three new *phenomenologies of the inside:* psychoanalysis, X-rays, and cinema. Three techniques or technologies that pursued the scene of interiority, the opening of the mind, the body, and the world. X-ray radiation was perhaps the emblem; the mysterious ray a figure for the body of a new form of light that yielded a new visuality, a modern form of light and its transmission that permeated the twentieth century. X-ray photography represented the advent of a new technique, one that explicitly recorded the destruction of its object, producing at once an optics and archive of annihilation. In question was a mode of writing and unwriting of disaster that moves from X-ray to atomic radiation and traverses the cinema. The scene of Hôichi's suspended visuality has its origins in 1895 as much as in 1945, when the force of radiation moved beyond closed rooms and cameras, into the world— when the light of atoms exposed the earth and the universes around it.

• • •

In the universal Library imagined by Jorge Luis Borges, every book ever written and those yet to be written—the possibility of every book—form a fantastic and final collection of "all that is given to express, in all languages." The Library of Babel represents the last archive, the archive at the end of expression, the end of the archive on the occasion of its completion. Final and finite, but also infinitely pointed toward the end, toward finitude, virtually final, *infinal.* "It suffices," says Borges, "that a book be possible for it to exist."[1] This possibility, like the translatability that Walter Benjamin imagines—"the translatability of linguistic creations ought to be considered even if men should prove unable to translate them"—allows the book to exist before its appearance, to haunt its existence and mark its finitude as a premature phantom.[2] The possibility of every book (every future book) already lies in the last archive. Every book haunted by its future. But not only books: the Library also comprises graphic inscriptions and traces, and "for every sensible line of straightforward statement, there are leagues of senseless cacophonies, vernal jumbles and incoherencies."[3] From the sensible to the nonsensical, the audible to the inaudible, the lyrical to the strained, all orders and disorders of language appear in the Library. Every trace of past and future is assembled in this one place, an archive of "all that is given" to expression.

> Everything: the minutely detailed history of the future, the archangels' autobiographies, the faithful catalogue of the Library, thousands and thousands of false catalogues, the demonstration of the fallacy of those catalogues, the demonstration of the fallacy of the true catalogue, the

Gnostic gospel of Basilides, the commentary on that gospel, the commentary on the commentary on that gospel, the true story of your death, the translation of every book in all languages, the interpolations of every book in all books.[4]

Every word written and unwritten, yet and never to be written, endlessly divided: a record of every event that has occurred, will occur, and will never occur. Future histories and histories of the future, minutely detailed and timeless, an archive of timeless time and history, of the untimely nature of both. Each moment divided into endless variations, infinitely. You are also here, in it, as a figure of the future; as is your future, your future death, which has been archived a priori. (You are in the universe, but you are also a figure of the universe itself.) Borges's archive includes, among everything, "the true story of your death." It includes the authentic narrative of your death (as well as, presumably, an infinite number of false accounts), a death that has already happened, will have (already) happened when it finally arrives. Perhaps it *has* already happened, and you are only an afterthought, its shadow, a trace of your (own) death, which is there already, as if another's.

In the vast archive of an infinitely divisible space and time, you are an atom: singular and indivisible: a "hypothetical body, so infinitely small as to be incapable of further division" *(Oxford English Dictionary).* "Atomism, the preeminent discourse of Western materialism (at least in antiquity and, more convincingly, since the seventeenth century)," says Daniel Tiffany, "contends that the authentic material existence of any body (as opposed to its merely phenomenal appearance) consists of infinitesimal and irreducible particles called atoms."[5] Because of the invisibility of atoms, atomism relies, says Tiffany, on images. Your death imagines you as a hypothesis. Atomic space emerges from a universe divided infinitely to the point of a hypothetical irreducibility. The logic of the atom, the *atomology,* sustains a paradox: infinite divisibility that ends, despite the endless nature of atomic division, in a hypothetical figure, small and indivisible, and also perhaps invisible. Gilles Deleuze calls this effect of "pure becoming" "the paradox of infinite identity."[6] You are an atom in Borges's Library, an infinite identity, infinitely divisible.

The Library of Babel exemplifies the promise of all archives, the fantasy and "fundamental law of the Library": totality. And so it "exists *ab aeterno,*" in and toward the future, always promised, imminent, here already but only ever as yet to come.[7] The totality of the archive effects an architectural temporality that is virtually universal. As a structure, an *arkhê,* it consumes almost all space, all future space, except for an irreducible sliver, which always remains as a surplus of the archive. In this small, indivisible, and atomic crevice in space and time, you reside, under the archive's shadow. This is the paradox of the Library of Babel, of its law, time, and fantasy: the universal Library takes all space, continues to expand, but always leaves space; its taking is a form of leaving, which casts, in the difference between totality and virtuality, a shadow. A shadow made possible only by the imaginary and hypothetical space always left vacant by the archive. A surplus space of images that can never be filled. You are there. You are the limit and surplus of the archive—an irreducible, atomic shadow of the archive.

Under the shadow, another form of life emerges, excluded from the archive that includes everything. "The certitude that everything has been written negates us or turns us into phantoms," says Borges.[8] Spectral because, in the archive's depths, your death has already been recorded and is no longer proper to you. Your death comes from and returns to the archive, its origin and destination. You are a phantom, a secret form of life, a life formed around a secret, around the pursuit and preservation of a secret. "When it was proclaimed that the Library contained all books, the first impression was one of extravagant happiness. All men felt themselves to be the masters of an intact and secret treasure."[9] The enormity of the secret, according to Borges, eventually turned this happiness to despair. "The certitude that some shelf in some hexagon held precious books and that these precious books were inaccessible, seemed almost intolerable."[10] Totality renders the world infinite, and thus distant: everything that belongs to you, that describes and reveals you, is far away, elsewhere, secret. The "intolerable certitude" can only manifest itself as a secret. At the heart of the archive, in the innermost chamber of its interiority, hides something that waits, like Kafka's law, only for you.[11] The certitude that the true story of your death exists somewhere in the archive, inaccessible, renders that account secret. A secret kept only from but also for you. An archive of secrets, of your secrets, which are at once singular and universal.

Imagining the universal archive produces an architectural impasse—a true Tower of Babel—since the archive requires all the space of the universe, which is still not enough. Its infinite form requires unlimited but finite, material space. An architecture suited for the Library of Babel must comprise universality, totality, and confusion. Something like an unconscious, where words are things, and things words, and the occupation of all available space (many times over) still leaves empty spaces. The universe, unconscious. In Borges's Library, the infinitude of language, of the possibilities of inscription (the written and yet to be written), determines the specificity of the archive, its architectonic law. (Language is, for Borges, the possibility of space.) The true archive fills space, all of space, and still needs more space; it becomes indistinguishable from space as such. Universal. "The universe (which others call the Library) is composed of an indefinite and perhaps infinite number of hexagonal galleries, with vast air shafts between, surrounded by very low railings."[12] Imaginary and material, "I prefer to dream," says Borges of the architecture, "that its polished surfaces represent and promise the infinite."[13] Those "indefinite and perhaps infinite number of hexagonal galleries" form a metonymy of space—the archive forms in space, is itself a form of space, casts a shadow in space, but is also a figure of space.

Despite Borges's careful attention to detail, his meticulous descriptions of hexagonal galleries, spiral stairways, and circular chambers, the Library remains, in some irreducible manner, formless. Division upon division, the archive appears like an atomic fissure without end. Every apparent limit yields more divisions, possibilities, and futures. In the end, Borges's description leaves the Library unimaginable. *The Library is a sphere whose exact center is any one of its hexagons and whose circumference is inaccessible.*[14] The surfaces and shapes, forms and lines that constitute the archive move steadily toward "fathomless air" and "inexhaustible stairways." An abyss, like the unconscious, always right there but elsewhere; nearby and material, yet always phantasmatic. In its material specificity, Borges's archive is formless and a figure of formlessness. A phantasm.

Borges's extreme architecture attempts to visualize the universe by assigning to every object real and unreal, now and yet to come, a code or sign, a corresponding figure within the Library. It seeks to render totality visible, to effect a total visibility and visuality. The Library of Babel is a view of the universe inside and out, an X-ray of the universe and universal X-ray, seen from within and without. It is a representation of everywhere: a perfect duplication of the universe. And of you: universal. An endless and eternal cinema, an imaginary archive that extends into the universe until it is indistinguishable from it, until you are indistinguishable from the universe.

Does the universe that Borges envisions, the fantastic archive that contains every inscription and future inscription, also contain the record of all that is not only uninscribed but uninscribable, every trace of the unrecordable and unrepresentable? And what form of trace does the uninscribable leave? Is the unwritable archived in the same fashion as the unwritten? Does it occupy space in the same way and present itself to the various regimes of representation in the same manner? Does an archive of the uninscribable exist apart from the Library of Babel, or is it the basis of its possibility, a secret architectural condition of the universal archive? Is the archive of the uninscribable, unwritable, and unrepresentable possible only as the destruction of the archive? Is it necessary to found, always in the ruins of the archive, a shadow archive? Another archive that replaces the archive, that takes place in its ruin as an afterthought and effect of destruction?

Which archive survives in the end? Which one remains, the archive or its shadow? What remains? On the survival of the archive, the universe, the "Library of Babel," Borges's narrator concludes: "I suspect that the human species—the unique species—is about to be extinguished, but the Library will endure: illuminated, solitary, infinite, perfectly motionless, equipped with precious volumes, useless, incorruptible, secret."[15] In the Library is the end of the world, the true story of the world's end. Between the finitude of the whole (the imminent extinction of the human species) and the singular (your death) opens an architecture of survival: an architecture that survives individuality and multiplicity. An architecture of the world's end. It remains, formed in space, measured and calculated in space, and yet in the end, in an imaginary end, formless, avisual, secret. In the infinite Library, the total archive that expands outward without end, the only limit is *you*, the limit that you encounter as your death, your proper limit. It is this sole limit that one discovers in the archive. You are its only limit, absolutely singular and universal, formless. You are its nucleus: indivisible and secret.

● ● ●

There, an archive burns. In the distance. Because the archive engenders an irreducible distance regardless of its proximity, the archive burns at a remove, always. There, it burns in a distinct manner peculiar to archives. It burns both internally and externally: engulfed by flames from the outside, passion and fever on the inside. A place at once interior and exposed, imaginary and material: a public interiority and secret exteriority. An archive is always there, just there, beyond this point, in flames. As all archives are destined to do, this archive burns, is constituted as archive by burning, by leaving the traces, or cinders, or remnants by which each archive is, in the end, constituted. In its own end. This and all archives are realized in destruction, preserved by the traces of destruction. Those traces or ruins of the archive are, in some fundamental way, neither internal nor external to the archive, neither inside nor outside it. Interiority and exteriority abandoned in the architecture and geography of the archive and its surroundings, the archive is the limit and the end of the limit between interiority and exteriority.

Because of its unique architecture, an archive burns simultaneously from the outside in and from the inside out. The origin of its destruction comes from within and without. In the distance and from the ruins of the archive emerges the figure of another archive, a secret archive, an archive of secrets, "the very ash of the archive." A secret archive constituted as secret on the occasion of the archive's destruction. A secret that precedes the archive and serves as its condition of possibility and impossibility. A secret reality. "Reality," Nicolas Abraham and Maria Torok say, "is defined as a *secret*."[16] From this reality, one then two archives. An other archive that is essential, without being any less foreign to the archive it replaces. This other archive takes place in the ruin of the archive, an event of the ruin. Other because, as Jacques Derrida says, there can be no archive of the secret *itself,* "by definition." No archive of the secret, which is, by definition, "the very ash of the archive."[17] ("A secret doesn't belong, it can never be said to be at home or in its place.")[18] "In an archive, there should not be any absolute dissociation, any heterogeneity or *secret* which could separate *(secernere),* or partition, in an absolute manner."[19] By definition, the archive is indivisible. In front of the archive, toward but also inside it, one stands always *before the law* of the archive, this law of the archive, and defined by it.[20] A law that calls forth the secret in order to banish it, but also, always, to inhabit it: the atomic law of *indivisibility.*

Undefined and forged in ash, the shadow archive, an "archive of the virtual," as Derrida calls it, erupts from the feverish imagination of a *mal d'archive,* an archive illness and desire that *burns with a passion.*[21] (An archive sickness in the sense of a lovesickness, the secret archive burns with a longing and fever that nurtures the irreducible distance.)[22] The archive burns always over there, beyond the limits and the law of the archive. In secret and passionately. Against this fever, the secret archive protects the distance, its interiority from infection, contagion, fire.

But a secret archive is not the same as an archive of secrets. Among various architectural theories of the secret, psychoanalysis makes possible one passage to such an archive of secrets by recognizing the essential heterogeneity of the secret. (Secrets, one can say, configure space heterogeneously.) If, as Derrida says, there can be no heterogeneity of the archive, no secret or separation, then the secret, defined by the heterogeneity it imposes, demands an other archive, another form of archivization that preserves the heterogeneity. Thus the other archive has to be recognized as irreducibly other, even if it retains a form of proximity and familiarity that would otherwise make it appear similar. (Heterogeneous with respect to the topology and the economies of visibility: the secret is never located entirely on the inside or outside, never entirely visible or invisible. Psychoanalysis can be said to function as a kind of technology of the inside, an apparatus of the invisible. It determines, throughout Freud's oeuvre, an archaeology of the secret. A science of the secret, a secret science.) This other technology of the archive, the archive of secrets, appears in the mechanism of repression, which is bound to the dialectics of consciousness, but also to memory and history, two distinct archives separated by the protocols of homogeneity.

A science of memory and commemoration, psychoanalysis forges an archaeology of repression and of history, bringing them together in an uneasy stasis. A heterogeneous archive of the memorial and historial. In *The Gift of Death*, Derrida says that "history never effaces what it buries; it always keeps within itself the secret of whatever it encrypts, the secret of its secret. This is a secret history of kept secrets."[23] Psychoanalysis brings together the flows of memory and history, the personal and impersonal, interior and exterior lines of demarcation: the elements that at once make a heterogeneous thinking possible, and an archive—by definition—impossible. Psychoanalysis, in Derrida's formulation, makes another archive possible, an alternate archive split against itself, against the very law of the archive, a shadow archive in the distance and alterity of the other. An outlaw or other archive, an archive of the other, an unconscious archive. An archive made possible by the heterogeneity instituted by repression and the secret. "As if one could not," Derrida writes, "recall and archive the very thing one represses, archive it while repressing it (because repression is an archivization), that is to say, to archive *otherwise*, to repress the archive while archiving the repression."[24] To archive *otherwise—anarchivize*.

The open archive *exposes*, it reveals outward. It orients itself toward the outside, forming a kind of public interiority, opening itself to and onto a public space. The other archive, the shadow or anarchive, represents an impossible task of the archive: to protect the secret, its heterogeneity, and divide the archive from itself. It is an archive that, in the very archival task of preserving, seeks to repress, efface, and destine its own interiority to oblivion. But to what end? And how does such an anarchive address the public from which it conceals? Giorgio Agamben, following Michel Foucault, formulates a definition of the archive that reflects the possibilities of enunciation within the vicissitudes of memory. Memory and oblivion, and the said, sayable, and never said determine for Agamben a private and universal archive of sorts: "Between the obsessive memory of tradition, which knows only what has been said, and the exaggerated thoughtlessness of oblivion, which cares only for what was never said, the archive is the unsaid or sayable inscribed in everything by virtue of being enunciated; it is the fragment of memory that is always forgotten in the act of saying 'I.'"[25] A universal archive formed around and upon an oblivion of the subject. (To this Derrida adds: "Once I speak I am never and no longer myself, alone and unique. It is a very strange contract—both paradoxical and terrifying—that binds infinite responsibility to silence and secrecy.")[26] Agamben's fragment of memory, lost at the moment of uttering "I," destines the word, torn from the subject in the moment of enunciation and immediately forgotten, to the archive of secrets, an unconscious of the spoken, the anarchive. "The archive is thus the mass of the non-semantic inscribed in every meaningful discourse as a function of enunciation; it is the dark margin encircling and limiting every concrete speech act."[27]

Agamben's "dark margin" of nonsemantic inscriptions encircles the other archive, an alternative archive outside the archive, anarchive and *par-archive*. A secret archive and an archive of secrets, this asemantic archive survives the archive's destruction. It is founded on its ruin, against the law of the archive as a paradox. It protects what will have been a secret, the topology of the "secret history of kept secrets." The other archive, shadow and anarchive, preserves the history that has never been a history, a history before history, destroyed, as it were, before becoming history as such. An archive of that which has not been—a universe of the unarchivable.

1. The Shadow Archive (A Secret Light)

> But of the secret itself, there can be no archive, by definition.
> The secret is the very ash of the archive . . .
> —*Jacques Derrida,* Archive Fever: A Freudian Impression

Sigmund Freud's *Moses and Monotheism: Three Essays* (1934–39) and Tanizaki Jun'ichirô's *In Praise of Shadows* (*In'ei raisan,* 1933–34) disclose two archives that were never constructed or, rather, that were constructed as virtual archives. They belong to an imaginary architecture. Almost or nearly archival, these structures name two virtual sciences, one Jewish, the other "Oriental." Imaginary, phantasmatic, racialized archives. From two ends of the twentieth century, these two archives speak to the question of the end, an end, one could say, of the twentieth century, what will have been its end—in cinders and ashes. Psychoanalysis, says Jacques Derrida, "aspires to be a general science of the archive, of everything that can happen to the economy of memory and to its substrates, traces, documents, in their supposedly psychical or technoprosthetic forms."[1] A virtual science, perhaps, of the shadow, a shadow science.

Moses and Monotheism, which serves in part as the occasion for Derrida's reflection on the archive, and more generally on psychoanalysis and

Freud's house in exile, begins with a secret. Freud's 1938 "Prefatory Note," written in Vienna before his own exodus, concludes with this determination: "I shall not give this work to the public."[2] He then adds:

> But that need not prevent my writing it. Especially as I have written it down already once, two years ago, so that I have only to revise it and attach it to the two essays that have preceded it. *It may then be preserved in conceal-ment till some day the time arrives when it may venture without some danger into the light,* or till someone who has reached the same conclusions and opinions can be told: "there was someone in darker times who thought the same as you!"[3]

Into the light without danger. Freud's image of light, a secret light of the future, sustains the possibility of refuge in visibility as well as the danger of exposure. His evocation of "darker times," spoken from a hypothetical future, suggests a future light, the light of the future—the future as a form of light—but also the temporalities of light. What is the temporal nature of light? What is the luminosity of the future? Both speculations (on the radiant future, and the futures of illumination) are bound by a relation, in Freud's secret preface, to inscription. To and toward inscription. Writing is destined to light, but to the uncertainty of safety or persecution. Freud ini-tiates an economy of writing in light (photography), of writing to enlighten darker times, and of consigning writing to darkness, to danger, to suppres-sion. Writing is here severed from giving, relegated instead to an alternate passage to the archive. Freud's refusal to give to the public and the secrecy to which he instead gives over *Moses and Monotheism* resonates with Der-rida's insights on the logic of the gift. In his reading of Jan Patočka, Derrida says of "the *gift that is not a present,* the gift that remains inaccessible, un-presentable, and as a consequence secret":

> The event of this gift would link the essence without essence to the gift of secrecy. For one might say that a gift that could be recognized as such in the light of day, a gift destined for recognition, would immediately annul itself. The gift is the secret itself, if the secret itself can be told. Secrecy is the last word of the gift which is the last word of the secret.[4]

By withholding his manuscript from the public, Freud initiates an econ-omy of the gift, a donation to the archive of secrets. A secret donation.

The spatiohistorical dimensions of Freud's dilemma are interwoven throughout the texts and parentheses that constitute his work: the interi-ority and exteriority of Judaism, Freud's life and the threats that assail it, the future of psychoanalysis, and Freud's book on Moses, itself a Mosaic, form the book's archival dimensions. The original thesis (a thesis of ori-gins) is described in "Moses an Egyptian," which Freud published in *Imago* in 1937. It represents the first essay of *Moses and Monotheism* and posits the Egyptian origins of Moses. Freud's thesis suggests the irreducible exteriority of Judaism, which comes from elsewhere, embodied by an Egyptian. Freud sees his assertion as a form of negation or erasure of essential Judaic propriety, a patricide. "To deprive a people of the man whom they take pride in as the greatest of their sons is not a thing to be gladly or carelessly undertaken, least of all by someone who is himself one of them."[5] (Freud has reversed the lineage, inscribing Moses now as the son of Jews; the patricide, then, is of the community of fathers.) The threat against him, another son of Judaism, and against psychoanalysis, in a way Freud's other paternity, comes from within, against "someone who is himself one of them."

After an etymology of the name of Moses, which proves, according to Freud, the Egyptian roots of the name, Freud offers a speculative genealogy of the individual: the secret conception of Moses to aristocratic parents, the prophecy warning of imminent danger to Moses's father, the infant Moses's condemnation to death at birth, exile upon the water in a casket, rescue by animals ("suckled by a female animal or by a humble woman"), return home, vengeance against his father, and "greatness and fame."[6] The very name and life of Moses suggest an uncertain and complex trajectory, which moves secretly from inside to out, outside to in. The lines that separate the spaces proper to Moses are obscured by Freud's account. The Egyptian origins of Moses disclose, for Freud, "a secret intention," the need to inscribe a narrative of heroism over the foreignness of Moses. What is hidden in traditional accounts is the radical exteriority of Moses in every respect. What is seen as inside, what gives the inside its essence, its law, comes from the outside, from an unlocatable outside, an outside whose exteriority has been absorbed by the phantasm of interiority. The ghost of the outside that haunts the structure of the inside, its constitution, but also its archive.

In the second essay of the Mosaic, "If Moses Was an Egyptian . . . ," published together with the first in *Imago*, Freud resumes the question of Moses's Egyptian roots with no apparent break. Or, the break in time between the two essays remains unmarked in the published essays, which are continuous. If the second installment begins without the representation of a spatial break, it nonetheless begins by breaking the promise with which Freud concludes the first essay: "It will therefore be better to leave unmentioned any further implications of the discovery that Moses was an Egyptian."[7] Yet Freud does not appear willing to let go.

Resuming his speculation on the irreducible foreignness of Moses, Freud adds to the exteriorities brought to Jewish tradition and practice by Moses monotheism and circumcision. Freud concludes his analysis by describing the murder of Moses and the benefit that befell Yahweh, "a violent and bloodthirsty" local deity, as a result of Moses's religious doctrine. "The god Yahweh had arrived at undeserved honour when, from the time of Kadesh onwards, he was credited with the deed of liberation which had been performed by Moses; but he had to pay heavily for his usurpation."[8] In a dialectic of divine light, Freud describes the shadow that Moses's god cast on the figure of Yahweh.

> The shadow of the god whose place he had taken became stronger than himself; by the end of the process of evolution, the nature of the forgotten god of Moses had come to light behind his own. No one can doubt that it was only the idea of this other god that enabled the people of Israel to survive all the blows of fate and that kept them alive to our own days.[9]

The end of Freud's narrative brings the history of Mosaic Judaism and the Jews to the present day. To the day when Freud writes, under threat of persecution from within and without, "to our own days." Moses, the foreigner, other, outsider, has brought to Judaism its interior essence: monotheism, circumcision, and a secret god. The other god, hidden and overshadowed, guides his people from elsewhere. Moses, the mosaic of Moses and the essays that constitute Freud's book, determines a secret archive, unspeakable and invisible, formless in its mosaic heterogeneity, which makes possible another history of Judaism. Freud, it seems, still wants to say more, but resists for the time being: "I no longer feel that I have the strength to do so."[10]

In his 1938 Vienna preface, Freud regains his strength and resumes the task. "With the audacity of one who has little or nothing to lose, I propose for a second time to break a well-grounded intention and to add to my two essays on Moses in *Imago*."[11] Nothing to lose or nothing to give. Everything already lost or given, Freud's tenacious audacity revolves around the question of the gift and its relation to loss. Freud's contribution to the speculative origins of Mosaic Judaism derives from a gift given when there is little or nothing left to lose. In the first of two prefaces to the third and final installment of *Moses and Monotheism*, "Moses, His People, and Monotheist Religion," Freud identifies his new persecutors, who now threaten him from without, "the external danger," he says, from within the space of the outside, the inside out. Noting the violent suppressions in Soviet Russia and Italy at some distance, the Catholic Church at home, and "the relapse into almost prehistoric barbarism" in Germany, Freud remarks of the Nazi expansion, "the new enemy, to whom we want to avoid of being service, is more dangerous than the old one with whom we have already learnt to come to terms."[12] The enemy, the danger, is within and without, surrounding the space of writing, but already within it. Freud writes this preface in the dark, a writing he will continue in secrecy, as it were, without publication. A private publicity, or public privacy, Freud's gift consists of withholding the gift from an other that already, even before the gift has been offered, refuses to accept it.

A concealed and secret gift that waits to move "without danger into the light," that waits for another day (some day other than the present day, "our present days," to which Jewish history has led) to return, like Moses, to its destination and home. This preface marks a temporality of the gift and of writing, a temporality that measures space by the shifts in light and darkness, inside and out. The gift not given, whose giving is postponed; a gift suspended in time, a law, that waits to arrive, defines the contours of a secret archive. Given to be seen, there, in some incontrovertible manner, but absolutely unseen. Invisible in its absolute secrecy. Freud's gift requires a form of forgetting or dissociation that cannot be in the end undone. It destines his authorship to the modes of secrecy and pseudonymy, and to cinders. On the gift and its relation to "radical forgetting," Derrida says: "The thought of this radical forgetting as thought of the gift should accord with a certain experience of the *trace* as *cinder* or *ashes*."[13]

Moses and Monotheism, the text and *ecotext* that Derrida locates in *Archive Fever,* determines a series of paired terms—Moses and monotheism, Father and Son, father and son, darkness and light, inside and outside, book and body, sacrifice and gift—that are bound together by an economy of the secret. From the vantage point established by Derrida's work, the contours of the other archive become visible, an archive that follows the passage of the gift toward annihilation, visible ultimately in its invisibility. For the gift to remain a gift, says Derrida, it must never be acknowledged, destroyed at once, and forgotten. "The simple identification of the gift seems to destroy it. . . . It is as if, between the event or the institution of the gift *as such* and its destruction, the difference were destined to be constantly annulled."[14] The gift's destruction leaves traces, like those of an outside Moses; like the shards of a fractured memory that pierce the mnemic archive. Destruction of the gift, of the past, of the archive: these acts are bound, from Moses to Freud to Derrida, by the secret architectonic of the archive.

Freud's gift of renunciation can be seen as architectural because it puts into place the structures of a secret, a topology of the archive, but also because it establishes the boundaries of the manuscript or book, the surfaces that separate interiority from exteriority. Freud consigns this manuscript of many disparate pieces—a virtual and infinite Mosaic—to darkness, but also to the outside.[15] "*There is no archive without a place of consignation, without a technique of repetition, and without a certain exteriority. No archive without outside.*"[16] Freud exiles this work to the outside, to a period of latency. "Concealed" and suppressed, Freud's *Moses and Monotheism* forms a cryptic archive, sealed, secret, and remote like the prehistories of Moses himself. *Moses and Monotheism* is thus, according to the archive's logic, destined to return from the outside, from the future to which it has been designated. It is this exteriority and deferral that inscribes the law of the archive. The exiled text, banished to the darkness, forms a "*prosthesis of the inside.*"[17] It is sent outward to protect the interior, establishing that interiority from the outside. *Moses and Monotheism* haunts from the outside and future the origin, the *arkhê* of the archive.

In concealing the manuscript, Freud seeks to protect it from the destructive forces that surround him in 1938, the Catholic Church and National Socialism. By June of the same year, however, Freud's secret archive was complete. Freud added a second "Prefatory Note" from London in June 1938. From his new house in exile, Freud decides to move outside, in and from the outside: "I venture to bring the last portion of my work before the public."[18] "There are no external obstacles remaining, or at least none to be frightened of," he adds.[19] But the frontier of this crisis has moved inside; the "external dangers" are now "*internal* difficulties."[20] The movement from external to internal threat in Freud's prefatory pages is made more difficult by Freud's own migration from Vienna to London. Like Moses, Freud is now a foreigner: "Here I now live, a welcome guest."[21] However welcome, Freud writes from elsewhere. The book has become *unheimlich*, far away from Freud's city and home where he had lived for seventy-eight years since childhood. The third essay finished, the manuscript now ready for publication from England, it was also irreversibly estranged from its author. He writes: "I feel uncertain in the face of my work; *I lack the consciousness of unity and of belonging together which should exist between an author and his work.*"[22] As if someone else has written this conclusion; another Freud—in London, in another house—has finished writing the Mosaic trilogy. He has destroyed the suspended gift, which now returns to him elsewhere, to another archive, *as if* to another. Freud has been forced to destroy his relationship to the manuscript, relinquish his authorship, to preserve it. It returns to him as the text of another, signed by another; the book arrives like a returned gift.

The account that Freud offers in the third essay of *Moses and Monotheism* describes the return of the Mosaic religion to those very people, the Jews, who had sought to annihilate it. Tracing a speculative history of the Mosaic belief to the religion of Aten, which initiated the movement toward monotheism in its worship of the sun god of On (Heliopolis), Freud suggests that the religion of Moses, from the beginning foreign, was summarily rejected and then, after a long dormancy or latency, came to be seen as the very essence of Judaism.

We confess the belief, therefore, that the idea of a single god, as well as the rejection of magically effective ceremonial and the stress upon ethical demands made in his name, were in fact Mosaic doctrines, to which no attention was paid to begin with, but which, after a long interval had lapsed, came into operation and eventually became permanently established. How are we to explain a delayed effect of this kind and where do we meet with a similar phenomenon?[23]

Shock and trauma form for Freud the analogy to Moses and his treatment at the hands of his followers. Resistance and discomfort drove Moses away; shame and guilt brought him back slowly, quietly, over the centuries. Of the abandonment of Moses and his religion by the Jews, Freud says, "all the tendentious efforts of later times failed to disguise this shameful fact."[24] "But," he insists, "the Mosaic religion had not vanished without a trace; some sort of memory of it had kept alive—a possibly obscured and distorted tradition."[25] Some sort of memory, a "possibly obscured and distorted" secret tradition.

A secret archive had been formed to keep alive what would have been destroyed otherwise. In secret, the Mosaic archive grew larger and stronger. The Mosaic religion had been deposited into crypts formed in obscured and distorted traditions, where they were kept latent until it was safe to "venture without danger into the light." "The remarkable fact with which we are confronted is, however, that these traditions, instead of becoming weaker with time, became more and more powerful in the course of centuries, forced their way into the later revisions of the official accounts and finally showed themselves strong enough to have a decisive influence on the thoughts and actions of the people."[26] In Freud's thought, Moses has been absorbed by the Judaic unconscious, where he continues to exert an influence from afar, from obscurity, from the privileged site of oblivion.

And it was this tradition of a great past which continued to operate (from the background, as it were), which gradually acquired more and more power of people's minds and which in the end succeeded in changing the god Yahweh into the Mosaic god and in re-awakening into life the religion of Moses that had been introduced and then abandoned long centuries before.[27]

"The determinants which made this outcome possible are for the moment," Freud concludes, "outside our knowledge."[28] The repressed Mosaic religion returns to Judaism from the outside as if it were already inside. From an inside experienced as outside and an outside experienced as inside. A forgotten outside, an imagined inside. The lines that separate inside from out have been lost, but more significantly, perhaps, those very spaces—inside and outside—have ceased to operate topographically. Moses returns to Judaism, as Freud's book, alien and uncanny, will one day come back to him. When *Moses and Monotheism* returns to Freud, it will be as a gift from another; as a gift that returns from the inside out. Moses and *Moses and Monotheism* represent archives destroyed and lost, reimagined and returned—a law and topography of destruction and return, the very story of Judaism itself.

To protect the archive, his book of books, the secret and true history of Moses, Freud has had to remove himself from it, has had to erase his relation to it. This erasure signals a shift in the economy of the archive. The destruction that threatens the archive changes from a desire to a drive; it no longer originates in an agency (Catholicism, Nazism) but rather in a force, the force of destruction itself. The fire that illuminates the darkness of the archive also destroys it. Preservation is made possible by destruction, by the figureless drive to destroy. The secret archive is bound by this paradox. The death drive, says Derrida, never leaves "any archives of its own. It destroys in advance its own archive," leaving only traces, ghosts. It "is above all *anarchivic*."[29] And yet "the archive is made possible by the death, aggression, and destruction drive, that is to say also by originary finitude and expropriation."[30] The archive is driven by destruction, by its relation to death, achieved in the finitude established by the drive. Freud's narrative about the unacknowledged murder of Moses, who introduced the Egyptian practice of circumcision to the Jews, remains overdetermined by a law of destruction.

The drive that at once destroys and preserves the archive also renders it spectral. "It is spectral *a priori*," says Derrida: "Neither present nor absent 'in the flesh,' neither visible nor invisible, a trace always referring to another whose eyes can never be met."[31] Neither corporeal nor ethereal, transparent nor opaque, the secret archive assumes the properties of a phantom, a shade. From the cinders of the archive, in its cooling embers, a shadow appears: a shadow of the archive, its impression, but also a phantom archive. An archive haunted by the archive. And haunting "implies places, a habitation, and always a haunted house."[32]

The very architecture of Freud's book is haunted. Despite the two false starts, the two prefaces, produced transnationally, and the lengthy third essay of the Moses trilogy, Freud is still not finished. In part 2 of "Moses, His People, and Monotheist Religion," Freud interrupts his extended speculation with another prefatory remark, an apology for the repetition about to follow. Freud describes his struggle to give up the theme of Moses, but, he concludes, "it tormented me like an unlaid ghost."[33] Alongside the apologies and explanations that Freud offers for the text's repetitive nature, with its various stages and pieces, is the claim that the book has exceeded the author, his imagination, and his will: like the foreign Moses that haunts Judaism, or the memory of Moses's god that haunts Yahweh. "Unluckily an author's creative power," he says, "does not always obey his will: the work proceeds as it can, and often presents itself to the author as something independent or even alien."[34] This book presents itself to the author, its source, as if it were alien, other; it gives itself (back) to the author from outside. Freud has allegorized, in his prefaces, the speculative history of Moses he offers in the book. A Moses who also returned, as if other, to his origin, to his destination, and yet has also had the traces of his otherness erased by the Judaic archive of latency and trauma. Moses haunts this story of fathers and sons, Freud and writing, the secret archive, history and fantasy, psychoanalysis and the world that surrounds and threatens it.

• • •

The archive, from the Greek *arkheion*, Derrida says, designates "a house, a domicile," an address that "shelters in itself [the] memory of the name *arkhê*."[35] The archive is first a building, a physical structure that carries within it the trace of a place. Tanizaki Jun'ichirô's reflections on illumination and architecture carry some traces, from another side of the twentieth century, of the exile that marks any relation to the archive. Tanizaki's ironic thesis on the Japanese house, *In Praise of Shadows,* represents a lament, an elegy for a Japanese architecture destroyed by illumination, by an electric force, which disperses the shadows that linger in and constitute the essence of the Japanese house. The work is haunted, like Freud's speculations on Moses, by an atmosphere of exile and the shadow of destruction.

Tanizaki's dilemma throughout *In Praise of Shadows* concerns the integration of electricity—appliances and wiring, but also heat, noise, and "excessive illumination"—into the surfaces of the traditional Japanese house. In his descriptions of practical solutions for reconciling tradition with modernization, Tanizaki incites an ironic polemic between Western and Japanese style and form, hygiene and visibility, luminosity and opacity, and the interiority and exteriority of the body, as well as the house. Various spaces of the house are measured by their effects on the human body, by the degree with which they absorb the body, forming an exoskeleton around it. Recognized as a discourse on architecture and the Japanese house, Tanizaki's treatise also provides an extended speculation on the Japanese body and the extent to which that body is indivisible from the architectural spaces that surround it, that have been, precisely, constructed to house it. Of the "Oriental" preference for a dark aesthetics, Tanizaki says, "Were it not for shadows, there would be no beauty":

Our ancestors made of woman an object inseparable from darkness, like lacquerware decorated in gold or mother-of-pearl. They hid as much of her as they could in shadows, concealing her arms and legs in the folds of long sleeves and skirts, so that one part and one only stood out—her face.[36]

Only the face of Tanizaki's ancestral woman stands apart from the darkness that envelops and hides her, but also constitutes her in and from the shadows. She is inseparable from the shadow, indivisible from the space that engulfs her. *In Praise of Shadows* explores the question of visibility and divisibility (as well as invisibility and indivisibility), and the intrinsic relation between the two concepts.

Among the reflections that Tanizaki offers on Japanese architecture is his discourse on the toilet, a space he qualifies as "perfection" (*risô*, ideal). The liminal space it occupies within and without the main body of the house, its proximity to the outside, and its shadowy illumination distinguish, in Tanizaki's discourse, the Japanese toilet as the last vestige of a vanishing architecture of the senses. Tanizaki's toilet, where the author speculates that countless haiku were born, surrounds the occupant with the textures of nature, enhancing what Tanizaki, following Natsume Sôseki, calls a "physiological delight." In this discourse, Tanizaki connects architecture with an architectonics of the body: the space of the toilet that surrounds the body serves as a type of skin, an exoskeletal surface that extends the body's limits outward. In the toilet that Tanizaki imagines, the body disappears, is absorbed into the architecture and environment determined by the liminality of the toilet. In the space made possible by the toilet, the human body and nature are fused: indistinguishable and indivisible. Inside and outside and neither within nor without the house, the body vanishes, forming an archive that spans from the innermost interiority of the body to the vast exteriority that surrounds it. The frame is both inside and out: "*Gestell*," says Martin Heidegger of his homonymic neologism, "is also the name for a skeleton."[37] In the toilet's intimate space, the place where the body performs its most essential activities, the body is lost. Lost in the exteriority of its most essential activities.

Tanizaki's polemic against Western toilets takes the form of a disdain for excessive illumination and its association with hygiene. The dialectic of shadow and light, "elegant" frigidity and "steamy" heat, in Tanizaki's elaboration turns to the idiom of cleanliness and sexuality, when he again invokes the figure of a woman. Of the sterile cleanliness of Western toilets, white and tiled, Tanizaki writes, "What need is there to remind us so forcefully of the issue of our own bodies. A beautiful woman, no matter how lovely her skin, would be considered indecent were she to show her bare buttocks or feet in the presence of others; and how very crude and tasteless to expose the toilet to such excessive illumination."[38] Tanizaki's use of the figure of a woman with beautiful skin who becomes obscene by revealing her "bare buttocks or feet" is itself revealing in the corporeality that sustains his discourse on what is perhaps the most corporeal space of the house. The toilet is anthropomorphic, gendered, and feminized in Tanizaki's imagination; its dark surface like the skin of a woman's body. Visibility makes the secret body indecent. Light is equated with indecency, enhanced visibility with grotesque visuality. The exposed female figure emerges in several key historical and rhetorical instances from the dream of psychoanalysis to the discovery of X-rays, determining a crucial relationship with tropes of interiority, invisibility, and the archive. Of the connection between excess visibility and sanitation, Tanizaki concludes: "The cleanliness of what can be seen only calls up the more clearly thoughts of what cannot be seen. In such places the distinction between the clean and the unclean is best left obscure, shrouded in a dusky haze."[39] Light and cleanliness, represented by the overexposed body, are opposed, in Tanizaki's logic, by the dark body, shaded, obscure, and unclean.

Tanizaki continues by extolling the aesthetic value of "the glow of grime." "If indeed 'elegance is frigid (following Saitô Ryokû, *fûryû wa samuki mono nari*),'" he says, "it can as well be described as filthy." The unclean generates beauty, an elegance proper to it, but also produces its own light, or glow.

> There is no denying, at any rate, that among the elements of the elegance in which we take so much delight is a measure of the unclean, the unsanitary. I suppose I shall sound terribly defensive if I say that Westerners attempt to expose every speck of grime and eradicate it, while we Orientals carefully preserve and even idealize it.[40]

Uncleanliness determines for Tanizaki a form of negative luminosity, but also an experience of temporality specific to the improper. "For better or for worse," he concludes, "we do love things that bear the marks of grime, soot, and weather, and we love the colors and the sheen that call to mind the past that made them."[41] Darkness and uncleanliness act as souvenirs of the past, as photographic mementos. They produce an impropriety of the subject, an uncleanliness that divides the subject from itself. In the revisionist logic of Tanizaki's theory of the unclean and improper, that which comes to the body from the past, the grime that accrues to the body over time, divides the body from its environment, but also comes to form the body as such: what is improper becomes proper, what is added to the body from the outside—from elsewhere, but also the past—becomes intrinsic to the body. The unclean body and the unclean spaces that surround and adhere to it come to represent an authentic state of the body as fundamentally improper. The elegant body is that which has been contaminated, a surface marked by the passage of time and the decay of matter, whose propriety is no longer discernible. Neither personal nor impersonal, the dark body that Tanizaki imagines emits an obscure light that destroys the lines between inside and out.

From private to public, a distinction destroyed by excess light, Tanizaki considers the Westernized architecture of twentieth-century Japan. Of the Western-style hotel, an archive of the other in modern Japan, Tanizaki complains about the lights and, more important, the heat they emit. "Worse than the waste," he says, "is the heat."[42] In the Miyako Hotel in Kyoto, Tanizaki describes "a white ceiling dotted with huge milk glass lights, each sending forth a blinding blaze."[43] "One can endure a Japanese room all the same, for ultimately the heat escapes through the walls. But in a Western-style hotel circulation is poor, and the floors, walls, and ceilings drink in the heat and throw it back from every direction with unbearable intensity."[44] The excessive illumination (here a Western imposition, *enlightenment*) drives away the shadows:

> As in most recent Western-style buildings, the ceilings are so low that one feels as if balls of fire [*hi no tama*] were blazing directly above one's head. "Hot" is no word for the effect, and the closer to the ceiling the worse it is—your head and neck and spine feel as if they were being roasted. One of these balls of fire alone would suffice to light the place, yet three or four blaze down from the ceiling, and there are smaller versions on the walls and pillars, serving no function but to eradicate every trace of shadow. And so the room is devoid of shadows.[45]

In the electric Western archive, in the *arkheion,* radiation descends from above and assails the body like a fever. It burns away the shadows, the virtual archive of an Oriental science. Total illumination. The blinding heat dispels for Tanizaki the phenomenon of an interior "visible darkness" (*akari ni terasareta yami,* an illuminated darkness), "where always something seemed to be flickering and shimmering, a darkness that on occasion held greater terrors than darkness out-of-doors."[46] The terror of shadows is replaced, in Tanizaki's account, by the blinding, burning force of light.

Tanizaki concludes his discourse on the house, the *arkheion,* by summoning another domicile, his proper dwelling; "the mansion called literature" (*bungaku to iu dendô,* a palace).

> I would call back at least for literature this world of shadows we are losing. In the mansion called literature I would have the eaves deep and the walls dark, I would push back into the shadows the things that come forward too clearly. I do not ask that this be done everywhere, but perhaps we may be

allowed at least one mansion where we can turn off the electric lights and see what it is like without them.[47]

Tanizaki conceives of literature as a residence, private and public, individual and national, imaginary and material—a fantastic mansion or archive. A shadow archive and an archive of shadows, the literary architectonic demands a resistance to excessive illumination. Against the drives of light and exposure, Tanizaki imagines a shadow archive, a literary archive of the Orient. To write literature, in Tanizaki's idiom, is to extend darkness, or at least to increase shadows, to introduce a visible darkness without light. A peculiar but precise logic permeates Tanizaki's discourse: to write is to expand darkness, to inscribe darkness, which forms in the end an archive. The archive is possible only as such a shadow architecture. Like Freud's mosaic thesis, itself a theory of the Orient and an Orientalist revision of Judaism, Tanizaki articulates a complex theory of writing and interiority, secrecy, and visuality. At the threshold of a limit, an abyss from which the very idiom of light would change irreversibly and forever, Freud and Tanizaki perform two acts of secret writing, two forms of shadow writing, which seek to protect two archives under assault in the 1930s.

In 1945 a Western force greater than electricity descended on the Japanese *arkheion*. The atomic assaults on Hiroshima and Nagasaki by the United States unleashed the heat and light of atoms, which threatened not only the Japanese archive but the "mansion called literature," the literary archive. It threatened to destroy the trace, to destroy even the shadows. The possibility of nuclear war, Derrida writes, "is obviously the possibility of an irreversible destruction, leaving no traces, of the juridico-literary archive—that is total destruction of the basis of literature and criticism."[48] Derrida consigns the nuclear war to the archive, literature, and the human habitat as an absolute referent. The fable of nuclear war (the story of a possible history yet to come) serves as the limit against which the archive survives: "If, according to a structuring hypothesis, a fantasy or a phantasm, nuclear war is equivalent to the total destruction of the archive, if not of the human habitat, it becomes the absolute referent, the horizon and condition of all others."[49] The *hypothetical* referent of a "total and remainderless destruction of the archive," Derrida insists, remains a fable, a fabulous fiction, which aligns it in some fundamental manner with literature, the *arkheion* or mansion of literature, the law of literature.

> If "literature" is the name we give to the body of texts whose existence, possibility, and significance are the most radically threatened, for the first and last time, by the nuclear catastrophe, that definition allows our thought to grasp the essence of literature, its radical precariousness and the radical form of its historicity.[50]

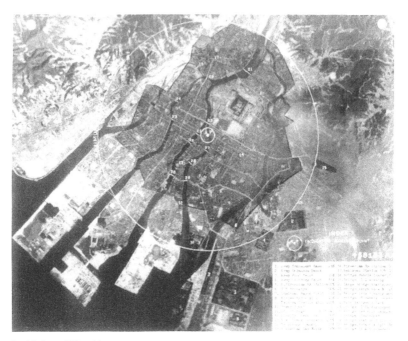

Aerial view of Hiroshima.

In this light, in the light of a searing Western heat, Tanizaki's anxiety over the vulnerability of the literary mansion, its fragility, appears prescient. Tanizaki trembles before a light still to come, under the first waves of catastrophic heat. A light that has already begun to appear, but has yet to reveal the full extent of its radiance. A radiation that arrives as atomic light and brings absolute destruction. Yet, as Derrida notes, the "explosion of American bombs in 1945 ended a 'classical,' conventional war; it did not set off a nuclear war."[51] The last war, which divided all wars from future wars, from the impossibility of war; the last conventional war was a divided war, no longer one, a war that can no longer be one. In the end, the mansion remained, remains, *manêre*, still stands and stands still in the smoldering embers of the archive. "Trace destined, like everything, to disappear from itself, as much in order to lose the way as to rekindle memory. The cinder is exact: because without trace it precisely traces more than an other, and the other trace(s)."[52] The "total and remainderless destruction of the archive" remains a *hypothesis* cast beneath the shadows of the archive.

When the smoke dissipates, the incinerated archive resurfaces, an archive in ruin, of ruins, of ashes. "*Il y a là cendre*," "cinders there are." From *Cinders:* "—What a difference between cinder and smoke: the latter apparently gets lost, and better still, without perceptible remainder, for it rises, it takes to the air, it is spirited away, sublimated *[subtilise et sublime]*. The cinder—

The Atomic Bomb Dome *(Genbaku dômu)*, originally known as the Hiroshima Prefectural Industrial Promotion Hall, was one of the few buildings to remain standing within a two-kilometer radius of ground zero.

falls, tires, lets go, more material since it fritters away its word; it is very divisible."[53] Derrida's distinction between smoke, which dissipates into the sky, and cinders, which fall to the ground, remaining divisible and material, illuminates a paradox, perhaps an irreconcilable contradiction in the visual economy of the concept and trope of cinders. Cinders are, in Derrida's idiom, what remains of a body that has vanished without a trace. "It is a trope that comes to take the place of everything that disappears without leaving an identifiable trace. . . . Everything is annihilated in the cinders. Cinders is the figure of that of which not even cinders remains in a certain way. There is nothing that remains of it."[54] Divisible, then, until nothing remains: cinders incite an absolute divisibility. Nothing remains, no trace whatsoever, except divisibility as such. When the object of incineration has disappeared without a trace, only the cinders, which disappear themselves without a trace, remain.

There, in the pyroprosthesis of the archive, are cinders. "*Il y a là cendre.*" "Nothing will have taken place but the place."[55] Of the place of the archive, its *Dasein*, Derrida poses the following question: "How are we to think of *there*? And this *taking place* or this *having a place* of the *arkhê*?" For Derrida, the topology of the archive, its geography, requires one to think and write atomically. "From this point on, a series of cleavages will incessantly divide every atom of our lexicon."[56] In the archive, or toward it, language becomes atomic—microscopic, *deconstructed,* splitting incessantly into near imperceptibility. Atomic writing produces another site of writing, another scene, space, or archive in which another writing, secret and divided, hides. A genre of the other trace, atomic writing moves only toward the indivisible point, the end of divisibility, toward an irreducible figure, the fantastic subject of atomic writing, "you." The "archiving trace," Derrida concludes, "its immanent divisibility, the possibility of its fission, [is] haunted from the origin."[57] Against the archiving trace emerges the atomic trace, or cinders, which threatens to reduce the immanent divisibility of the trace to an irreducible, indivisible, invisible afterimage: it threatens an atomic reaction that will incinerate the archive.

> —The fire: what one cannot extinguish in this trace among others that is a cinder. Memory or oblivion, as you wish, but of the fire, trait that still relates to the burning. No doubt the fire has withdrawn, the conflagration has been subdued, but if cinder there is, it is because the fire remains in retreat. By its retreat it still feigns having abandoned the terrain. It still camouflages, it disguises itself, beneath the multiplicity, the dust, the makeup powder, the insistent pharmakon of a plural body that no longer belongs to itself—not to remain nearby itself, not to belong to itself, there is the essence of the cinder, its cinder itself.[58]

"A plural body that no longer belongs to itself," says Derrida. The other body—divided from itself—is defined by the fire, by the shadow it casts on the body, by the cinders it leaves on the body's surface, on its skin. The fire retreats into the shadow, hides only to retrace the path of the secret destruction of a virtual archive carried on the body, a mnemic archive inscribed on the surface of the skin—burned, as it were, on the skin's surface.

The pellicular surfaces—skin and film—yield an archive marked on the body. In the Freudian archive, Derrida discovers "a *private inscription*" on a book given to Freud from his father. The inscription is made on the book's skin, on the skin that constitutes the book's pages. Derrida says: "A very singular monument, it is also the document of an archive. In a reiterated manner, it leaves the trace of an incision *right on* the skin."[59] The private archive that moves in this instance from father to son in the form of a gift takes place on the surface of the book and the skin:

> The foliaceous stratification, the pellicular superimposition of these cutaneous marks seems to defy analysis. It accumulates so many sedimented archives, some of which are written right on the epidermis of a body proper, others on the substrate of an "exterior" body. Each layer here seems to gape slightly, as the lips of a wound, permitting glimpses of the abyssal possibility of another depth destined for archaeological excavation.[60]

The private inscription elicits an archaeology of the surface. To dig deeper, to excavate, is to return to the surface not as the inability to probe depths but rather as the capacity to render the abyssal features of the surface. A prehistory of the surface, an ancient history of the moment, of each moment at that moment. What remains in the destruction of the archive returns always to the surface, to the skin, as a skin, a remainder etched onto the skin, a book. A "memory," Derrida calls it, "without memory of a mark."[61]

The memories of the mark archived in the act of marking are also violent and destructive, registered in this case, in the gift from father to son. The relation of father to son, genetic and theologic, is bound by the gift of death, by the impossible economy of sacrifice. Genetic, instinctive, but also profoundly psychological, the impulse of the gift appears, according to Derrida, in Freud's notion of the drive, specifically the death drive. This drive, the most interior of interior conditions, leaves its mark on the skin, says Derrida; it forms on the body a set of inscriptions that turns it, like Hôichi's body, into a book, an archive of private inscriptions. Of the impressions left by Freud's death drive, Derrida remarks: "This impression of erogenous color draws a mask right on the skin.... These impressions are perhaps the very origin of what is so obscurely called the beauty of the beautiful. As memories of death."[62] Such memories of death linger on the skin as a form of exteriorized secrets. For Derrida, memories return to the body, to its surfaces and depths, like a gift—this particular gift, this book—from death.[63] They return on and below the body's surfaces as moving impressions, as a mobile affectivity. "A secret," says Derrida, "always *makes* you tremble ... a certain irrepressible agitation of the body, the uncontrollable instability of its members or of the substance of the skin or muscles."[64] The impressions to which Derrida alludes, always secret, always inscribed in the nonsemantic script of the secret, are nonetheless "*scriptural* or *typographic.*"[65] Drawn on the skin, inscribed yet cryptic, a *scrypt.* Derrida's spectral topology of the archive makes possible the return to two other archives, two archives of the other, two shadow archives that appeared in the same phantasmatic year as psychoanalysis.

● ● ●

From cinders and skin to shadows and impressions, another archive of surfaces emerges in the archaeology of the surface, of the archive, the secret archive of surfaces. In 1895, the year that Freud and Josef Breuer published *Studies on Hysteria,* photography crossed the threshold of the human body with the penetrating light of the X-ray. Referred to alternately as skiagraphs, photographs in reverse, or shadow scripts, X-rays exposed the secrets of the body, its depths, collapsing the essential divide between surface and depth, and rendering the body a deep surface. Also in 1895, a set of institutional, economic, and technological conditions now known as cinema emerged. Animated and projected photographs compressed the volume of life onto a screen, effecting an animated, photographic, and electric view that maintained, through the apparatus of the lens, a flat three-dimensionality. On the surface, on surfaces, X-rays and cinema introduced two forms of radical *photography,* two modes of writing with light that established—almost at once—new archives marked by a profound *superficiality.* Both graphic systems offered extreme, even excessive modes of visuality that came to be seen, paradoxically, as modes of invisibility, or unseeability, challenging the notion of interiority, of envisioning and probing interiority, but also the conditions of visuality as such. Each archive became, at the moment of its appearance, an antiarchive, an anarchive, a secret archive of the visible.

X-rays and cinema, along with the technique of psychoanalysis, established in 1895 new technologies for visualizing the inside, for imagining interiority; but they also transformed the conditions of visuality as such. Among the many effects of this reconfiguration of the inside and out, surface and depth, visuality and avisuality was the formation of a secret visuality. "Since things and my body are made of the same stuff," says Maurice Merleau-Ponty, "vision must take place in them; their manifest visibility must be repeated in the body by a *secret visibility.*"[66] The contiguity of the body and things, fused by the "same stuff," forms inside Merleau-Ponty's imaginary body a regime of secret visibility intrinsic to the body. X-ray, cinema, psychoanalysis provided a view of the secret visibility, not an access or opening as such, but a mode of avisuality. Those new phenomenologies introduced an alien you, secret and distant in its proximity to you. There, on the surface, but alien. A secret visuality of you as another, of you, who is here, in this image, pictured in the frame of a singular interiority, elsewhere and other. You are photographed—discovered and exposed—in the economy of a secret visuality. Seen in secret, seen secretly, a scene of secrecy.

Of secret visuality, Derrida asks, "To see in secret—what can that mean?"[67] A secret optics? A visible spectrum no longer material and universal but prone to the peculiar economies of the subject, desire, and secrecy? Derrida imagines a secret kept from the visible world, a secret whose secrecy is maintained by the other senses.

> One might imagine a secret that could only be penetrated or traversed, undone or opened as a secret, by hearing, or one that would only allow itself to be touched or felt, precisely because in that way it would escape the gaze or be invisible, or indeed because what was visible in it would

keep secret the secret that wasn't visible. One can always reveal to the gaze something that still remains secret because its secret is accessible only to the senses other than sight.[68]

The archive is formed here in the other senses; an archive of the visible that remains secret, a visible secret, a secret of the visible. Supplemental, the archive of secret visuality appears in the other senses, everywhere but in visuality. (Merleau-Ponty calls this an "echo.")[69] Derrida's secret is here maintained only under the regime of the visible from which it is excluded. ("The secret that is for me is what I can't see.")[70] What is seen can remain nonetheless invisible. What is nonetheless available through a system of the archive can remain invisible. Such archives leave traces, shadows, remnants in lieu of visible documents. "That which is hidden, as that which remains inaccessible to the eye or hand, is not necessarily encrypted in the derivative senses of that word—ciphered, coded, to be interpreted—in contrast to being hidden in the shadows."[71] To remain in the shadows, hidden, does not necessarily imply encryption, unless the very nature of the shadow is seen not as the negation of light but as the property of light and visuality that embodies its own positive attributes. A shadow materiality that illuminates a material darkness.

In *The Gift of Death*, Derrida describes two modes of invisibility, two ways in which the invisible appears. The first he designates with a hyphen, "in-visibility." Of this first order of invisibility, Derrida says: "There is a visible in-visible, an invisible order of the visible that I can keep secret by keeping out of sight. This invisible can be artificially kept from sight while remaining within what one can call exteriority."[72] Secret and outside, invisible, says Derrida, because kept out of sight. He offers two examples—one atomic, one corporeal—in an extended parenthesis.

(If I hide a nuclear arsenal in underground silos or hide explosives in a cache, there is a visible surface involved; and if I hide a part of my body under clothes or a veil, it is a matter of concealing one beneath another; whatever one conceals in this way becomes invisible but remains within the order of visibility; it remains constitutively visible. In the same way but according to a different structure, what one calls the interior organ of the body—my heart, my kidneys, my blood, my brain—are naturally said to be invisible, but they are still of the order of visibility: an operation or an accident can expose them or bring them to the surface; their interiority is provisional and bringing their invisibility into view is something that can be proposed or promised.)[73]

The body's interiority is provisional, like a nuclear arsenal hidden underground, in the earth. Although each is hidden from sight, they still belong to the order of the visible. In each case, the visibility is there but hidden, secret. "All that," Derrida concludes, "is of the order of the visible in-visible."[74] Derrida's figure for in-visibility moves from the atomic weapon to the body's interior; both can be exposed by an "operation or accident." In this light, according to the logic of in-visibility, the X-ray or skiagraph can be seen as an apparatus, a technique and technology, that seeks to render the in-visible visible, to expose the body's provisional interiority, but also to view the in-visible without disturbing its secrecy. X-rays record only the shadows of a secret, its trace, the place where it hides. Not so much an ex-posure as a disclosure, the X-ray reveals secret visibility as a mode of secret visuality, showing what nonetheless remains invisible, without operation or accident.

The second order of invisibility Derrida calls "absolute invisibility." "But there is also absolute invisibility, the absolutely non-visible that refers to whatever falls outside of the register of sight, namely the sonorous, the musical, the vocal or phonic (and hence the phonological or discursive in the strict sense), but also tactile and odoriferous."[75] This order of invisibil-ity is never given to sight; it resides in the other senses as invisible. It is, in a phenomenological sense, absolutely outside vision. Absolute invisibility establishes a form of secrecy preserved in the other senses. A visibility that takes place elsewhere, outside, but still, in some manner, visible, even as nonvisible. That is, even absolute invisibility or nonvisibility remains a form of secret visibility: it is seen in the other senses, as another sense. To see in another register, to hear or smell an image, to touch it. So *it is given* to be seen, only somewhere else. To see otherwise, to see in another sense, to see in secret.

Derrida's orders of visibility extend, in both instances, "beyond the vis-ible."[76] The in-visible withdraws from visibility without disappearing from the realm of the visible world, like a hand that conceals itself under the table: "My hand is visible as such but I can render it invisible."[77] The other, absolute invisibility falls entirely outside the "register of sight." Together, Derrida's excess visualities might point to a category of complex visuality, a system of visuality that shows nothing, shows in the very place of the visi-ble, something else: *avisuality*. Avisuality not as a form of invisibility, in the sense of an absent or negated visibility: not as the antithesis of the visible but as a specific mode of impossible, unimaginable visuality. Presented to vision, there to be seen, the avisual image remains, in a profoundly ir-reducible manner, unseen. Or rather, it determines an experience of seeing, a sense of the visual, without ever offering an image. A visuality without images, an unimaginable visuality, and images without visuality, avisuality. All signs lead to a view, but at its destination, nothing is seen. What is seen is this absence, the materiality of an avisual form or body. Like Derrida's

The Atomic Bomb Dome.

hand rendered invisible; like Freud's secret manuscript withdrawn from view, from the scene of writing; like Tanizaki's Oriental body, lost as it were, in the shadow archives of the Japanese house. This is a visuality that *moves away from view*, a cinema, and one that *burns away from view*, cinders. Bound by the secret homonym and visuality formed from two romanized classical prefixes, *cine*. A secret sign and homonym that signifies in two registers, bringing together ashes and movement. To *cinefy*: to make move, to make cinema and to incinerate, to reduce to ashes. Traces and residues of movement, and the movement of ashes. *Cinefaction*. Like the true story of your death, a secret cinema, registered in the unimaginable depths and interiorities of the Library of Babel at the end of the universe. Burning and disappearing, atomic...

▼ The Shadow Archive

2. Modes of Avisuality:
Psychoanalysis–X-ray–Cinema

Sigmund Freud's dream of visualizing the unconscious, his wish to archive the unconscious as a psychosemiography, came to him in late July 1895. It came as an actual dream; a dream carried by a dream that collapsed the figurative and literal dimensions of the word. In Freud's thesis on dreams, which he developed from his dream of 1895, this and every dream represents a continuation of thought, an unconscious wish. The unconscious, itself an archive of the human subject, comprises for Freud the material and immaterial traces of each individual; the history and pre-history of the subject and all the others that have contributed to the formation of each subject. An archive of the self, of the interiorities and exteriorities that constitute each individual. And like Borges's Library, the unconscious carries within its vast collection "the true story of your death," the narrative or drive that leads to your death, to the secret place of your death.

Following a catastrophic sequence of misdiagnoses earlier in the year that had almost resulted in the death of Emma Eckstein, Freud, his confidence shaken, began to question the viability of a medical practice that took as its object the immaterial form of the psyche.[1] The crisis had reopened in Freud a primal concern: can the psyche be the source of illnesses that manifest themselves on the body? What phantasmatic tissues suture the psychic and organic bodies? The crisis had threatened the very foundation of Freud's theory of repression and had caused him to rethink the possibility of a materiality and visuality of the psyche, a kind of psychic corporeality.[2]

During the night of 23–24 July 1895, a woman whom Freud names "Irma" appeared before him with the "solution" to his crisis. The "dream of Irma's injection," which Freud would later claim had revealed to him "the secret of dreams," marks a critical moment in the history of psychoanalysis: it was the first of his own dreams that Freud openly submitted to full analysis.[3] Against the uncertainties that assailed his nascent theory of the unconscious, Freud felt the pressing need to offer a material figure or image of the psychic apparatus. The dreamwork, constituted by signifiers of visuality, promised such a figure. The "Irma dream" suggested the possibility of a virtually impossible spectacle; an opportunity to observe the psychic apparatus in motion, acted out, as it were, on a dream stage by dream actors. Freud offers the following account of his dream, which he transcribed and analyzed in *The Interpretation of Dreams* (1900):

> A large hall—numerous guests, whom we were receiving.—Among them was Irma. I at once took her on one side, as though to answer her letter and to reproach her for not having accepted my "solution" yet. I said to her: "If you still get pains, it's really your own fault." She replied: "If you only knew what pains I've got now in my throat and stomach and abdomen—it's choking me"—I was alarmed and looked at her. She looked pale and puffy. I thought to myself that after all I must be missing some organic trouble. I took her to the window and looked down her throat...[4]

The dream stage opens onto a cavernous hall, an architecture that engulfs, like an oral cavity, Freud's cast. Among them is Irma, whom Freud takes to the side and reproaches. Irma has refused Freud's solution, his diagnosis.[5] Alarmed by her appearance, "pale and puffy," Freud brings Irma under the light of a window and peers into her throat and body, into her depths that open into another space, another cavern, an architecture of the body within the dream architecture. The interior design of Freud's dream architecture includes a passage to the outside. The dream window (there are several important windows in Freud's dream archive) represents a source of light and exteriority, an inside exteriority, an exteriority imagined from the inside.[6] The window provides a source of artificial light that masquerades as natural light, a theatrical light not unlike those recessed lights imagined by Tanizaki.[7] Irma resists Freud's probe, "like women," he says, "with artificial dentures." But she relents:

—She then opened her mouth properly and on the right side I found a big white patch; at another place I saw extensive whitish grey scabs upon some remarkable curly structures which were evidently modelled on the turbinal bones of the nose.[8]

A new space opens within Freud's dream, effecting a *mise-en-abîme* of spaces that unfold inside other spaces. From the inside of his psychic space to the inside of an imaginary architecture, from inside the cavernous hall to the inside of Irma's mouth, Freud moves deeper into the dream space. With the penetration of Irma's body—the dream will ultimately induce an unsanitary injection—Freud's figures become increasingly abstract. In the recesses of Irma's body, Freud observes abstract forms: "A big white patch . . . whitish grey scabs . . . curly structures which were evidently modelled on the turbinal bones of the nose." As he moves closer to the source of the pain that assails Irma, Freud's vision begins to dissolve and becomes increasingly formless. As if to establish a base of witnesses, and driven still by the impulses of wish fulfillment that he describes in *The Interpretation of Dreams*, Freud summons other observers to the scene of Irma's interiority.

—I at once summoned Dr. M., and he repeated the examination and confirmed it. . . . Dr. M. looked quite different from usual; he was very pale, he walked with a limp and his chin was clean-shaven. . . . My friend Otto was now standing beside her as well, and my friend Leopold was percussing her through her bodice and saying: "She has a dull area low down on the left." He also indicated that a portion of the skin on the left shoulder was infiltrated. (I noticed this, just as he did, in spite of her dress.)[9]

Perhaps to protect himself, to protect the desire and anxiety that erupt in his dreamwork, Freud assembles an audience of male physicians who confirm his analysis. In this scene, Freud enacts a second penetration. Both Leopold and Freud are able to identify a "portion of the skin on the left shoulder" that had been "infiltrated." This, Freud adds, "in spite of her dress." Freud's X-ray vision locates another opening on Irma's body: a secret orifice hidden beneath her dress, an infiltration, a secret passage to Irma's interiority.[10] It requires a form of perception not yet known, the ability to see through opaque objects and into the body's depths. For Freud it represents perhaps the kind of probing visuality made possible by psychoanalysis.

The dream concludes.

M. said: "There's no doubt it's an infection, but no matter; dysentery will supervene and the toxin will be eliminated.". . . We were directly aware, too, of the origin of her infection. Not long before, when she was feeling unwell, my friend Otto had given her an injection of a preparation of propyl, propyls . . .

propionic acid . . . trimethylamin (and I saw before me the formula for this printed in heavy type). . . . Injections of that sort ought not to be made so thoughtlessly. . . . And probably the syringe had not been clean.[11]

The origin of Irma's pain is an infection, which had developed after Otto had given her an injection with an unclean syringe. This improper injection had introduced into Irma's body a toxin. Freud's dream witnesses attest to this, relieving Freud of his responsibility, although it still seems that he had missed some "organic trouble." Freud concludes in the analysis of his dream that follows that the dream represented the fulfillment of a wish: the wish to be absolved of any responsibility for the continuing illness of Irma.

Freud's self-analysis, eccentric and frequently blinded, offers further insight into the crisis of visuality and imagination provoked by the dream of Irma's injection. Regarding Irma's symptoms, Freud swiftly reaches an impasse. "Pains in the throat and abdomen and constriction of the throat played scarcely any part in her illness. I wondered why I decided upon this choice of symptoms in the dream, but could not think of any explanation at the moment."[12] Freud has substituted Irma's most common symptoms—"feelings of nausea and disgust"—for the dream symptom of a choking pain. Irma is choking, in Freud's dream, unable to articulate or communicate a thought that seems to erupt, like nausea and disgust, from her stomach and throat. Freud has replaced Irma's nausea with a constriction of the throat. But he has not eliminated the nausea and disgust, which will eventually leave the body in dysentery. Using the same techniques that Freud himself developed and applied to dream analysis, one can speculate that Irma wants to say something to Freud who has, in the dreamwork, choked her, and forced back inside of her an accusation that threatens to ruin him. It is perhaps the expression of Irma's feelings of nausea and disgust for him, for his solution, which Freud is preventing.

At work, also by Freud's own admission, is a prominent feature of the dreamwork, displacement. Of Irma's "pale and puffy" appearance, Freud notes, "My patient always had a rosy complexion. I began to suspect that someone else was being substituted for her."[13] Freud ends his speculation for the moment, leaving open the possibility that this someone else might be an idea generated by the dreamwork, that is, not a person at all.[14] Irma's pale and puffy exteriority continues inside her mouth, where Freud sees whitish and grey scabs and patches. A skeletal continuum seems to have opened in Freud's imagination, from "artificial dentures"—the exposed bones that line the entrance into the oral cavity—to the "turbinal bones of the nose." Irma's demeanor, "like women with artificial dentures," her pale, puffy appearance, and the white oral patches that resemble the bones of the nose, but actually those of the female sexual organs, Freud later adds, form in Freud's scene an image of Irma inside out: Irma displays her interiority on the outside, she is *exposed*. She represents a displaced image or figure, but even more an image of radical displacement from depth to surface, inside to out. Irma's secret orifices, the direct passages to her interiority, are on her body, elsewhere.

Freud's dream is sustained by a unique architectonics, a structure of visuality that seeks to found concrete archives of the mind, body, and dream. The dream of Irma's injection, which came to Freud while he was vacationing at the Schloss Bellevue, "the beautiful view," near Vienna, can be seen as a view, a vision, a speculation that resists specularity. A wish perhaps for a view, a glimpse, an image of that which would secure the viability and visibility of psychoanalysis. Freud seems to have dreamed of visualizing the psychic form of Irma's disorder, imagining a psychography of the unconscious. He has attempted to secure the unconscious in the visual field of the most exemplary of visual indexes, the human body. This nervous fantasy haunted Freud throughout his career.[15] Only the dilemma is clear: the anticipated birth of psychoanalysis (Freud had already imagined a future plaque to mark the site of the dream) had been complicated by the return of a repressed corporeality. Organic and psychic disturbances were vying for supremacy in Freud's own turbulent psyche: "I thought to myself that after all I must be missing some organic trouble."[16] In the dream, Irma's interiority appears to obstruct Freud's view of the unconscious. Or rather, two competing forces are at work in the dream: one brings the image closer, by pushing deeper and further into the body, toward greater clarity, figuration, and materiality; the other follows the same route toward dissolution, disappearance, and formlessness. By compressing Irma's interiority onto the surface of his psyche, Freud has invoked a spectral architecture, an architectonic of interiority that moves at once toward representation and abstraction. Neither corporeal nor imaginary, the visuality of Freud's dream introduces, according to Jacques Lacan, a paradox: it generates a formless image, an image of formlessness. Lacan writes:

> There's a horrendous discovery here, that of the flesh one never sees, the
> foundation of things, the other side of the head, of the face, the secretory
> glands *par excellence,* the flesh from which everything exudes, at the very
> heart of the mystery, the flesh in as much as it is suffering, is formless, in as
> much as its form in itself is something which provokes anxiety. Spectre of
> anxiety, identification of anxiety, the final revelation of *you are this—You
> are this, which is so far from you, this which is the ultimate formlessness.*[17]

You are this, this form, which is so far from you, here. Lacan's specter of anxiety hovers in the space opened between proximity and distance, the self and another, another that returns in the form of an other self. Foreign and formless, barely recognizable. The other that returns here is formless, as are, in Freud's thought, the unconscious wishes that charge the dream-work. But this is you, and you are this: what is formless is the space that opens onto "the other side of the head," a phantom space and architecture that takes place atopically, founding a semiology of the avisual.[18] Of another spectral woman and the space of alterity she emits, Trinh T. Minh-ha writes, "The space offered is not that of an object brought to visibility, but that of the very invisibility of the invisible within the visible."[19]

The struggle between competing forces "of the flesh that one never sees," which results in formlessness, might be understood not as a dialectic between figure and abstraction but rather as the advent of a paradoxical graphic that superimposes the concrete and abstract elements of the unconscious into a representation of formlessness, a view of the unseeable, a picture of that which cannot be seen. Figurative and abstract, avisual. Freud is striving not toward the mastery of a complex graphic order but, rather, toward the possibility of expanding the thresholds of the graphic spectrum to include that which resists graphicality as such: of designating in the unconscious a visuality of the avisual, a visual order of the formless. For Freud, the semiotic trajectory of the dreamwork determines a phantom architectonics: a cartography of nowhere, an architecture of *nothing* (or the unconscious), and an archaeology of imaginary depth that always takes place on the surface. As a practice and sensibility, psychoanalysis remains attuned to superficiality; it constitutes a search for depth on the surface of things.

In the dream of Irma's injection, Freud discovers another opening on Irma's body, a secret and secondary orifice he locates in an act of extra-sensory perception through Irma's dress. Peering into the hole produced by a puncture, Freud discovers in this depth a formless *you*, an atopic and avisual subject. It is not Irma's unconscious that returns the look from in-side, but rather Freud's own. This is Lacan's "horrendous discovery." The locus of Irma's dream body provides the opening for Freud to encounter himself, the site of his own subjectivity deposited and reflected, as it were, in Irma's depths. He has himself infiltrated her, his desire the toxin. Freud is himself both the injection and the infection. What he finds in Irma's secret orifice is his own displaced unconscious incorporated by Irma's dream body. Her body becomes the vehicle for this displacement: "I began to suspect that someone else was being substituted for her." The fantasy of searching another's body for the residues of oneself articulates a wish that carries, as do all unconscious ideas, its antithesis: the desire to lose oneself in the other's body, to be dissolved in the body of the other. The discourse of that wish, Lacan argues, points toward a profound *senselessness*, a loss of meaning but also of one's senses in the cavernous opening of an abyss: "It is just when the world of the dreamer is plunged into the greatest imaginary chaos...that the subject decomposes and disappears. In this dream there's the recognition of the fundamentally acephalic character of the subject, beyond a given point."[20] Freud's apparent wish for such dissolutions would be fulfilled, within four months of his Bellevue dream, as a photographic phenomenon: the X-ray image.

Freud's dream prefigures the X-ray in coincidental and complex ways. When Freud looks, or dreams of himself looking into Irma, looking through her clothing at a puncture in her skin, he invokes the penetrating force of X-ray vision. He has discovered a secret passage to Irma's interiority, virtually visible and virtually invisible, made visible at all through an apparatus of extraperception. He also inscribes Irma's dream body in the arena of a medical mise-en-scène, instituting a spectatorship driven by the gaze of medicine and science. Freud's witnesses undoubtedly protect him from the eros that threatens to erupt throughout the dream of Irma's injection. The dirty syringe that penetrates the recently widowed patient; the injection that infects Irma, who suffered in her dealings with Freud from feelings of nausea and disgust; the economy of sexual allusions and references that sustains Freud's dream; and Freud's desire to expose Irma, to force her to yield her interiority—all are ensured by the medical authority that Freud summons in his dream. An authority that Freud lacked in his attempts to think and practice psychoanalysis. In this dream, Freud finds recourse in a medical scene that validates his credibility but hinders his wish to leave behind the organic residues of psychoanalysis, the constraints of conventional medical views. Like the X-ray image, Freud's dream confuses art and science, fantasy and fact, imagination and observation. Like psychoanalysis itself. The lines between those orders have been blurred, lost in an excess of the senses that results in a form of acephalia or senselessness. Interiority and exteriority, form and its absence, psyche and body, psychoanalysis and medicine, and art and science converge in this dream of dreams, the *arkhé-*dream, in which the secret of dreams—its archaeology—was revealed to Freud. The secret of dreams is also, in the final analysis, a secret dream.

The greatest prescience of Freud's dream concerns perhaps the form of visuality unleashed by the X-ray image. Irma's interiority opens to Freud not in the form of an unraveling that exposes what is hidden, that makes visible what is secret, but rather by making visible the "very invisibility of the invisible within the visible," as Trinh says. What is given to see is the unseeable, the shock of finding at the body's center a profound avisuality. Not an invisibility but an avisuality because something is given in the form of an excess. Its form, or force of visuality, exceeds the capacity of the spectator to see it, to withstand its very specularity. (This is also the unique visuality of the dream itself: dreams are visual but not visible; they are avisual images.) In this sense sublime: before the spectacular view, one is always lost, elsewhere, headless, and senseless. Freud's desire to transgress the body's surface, to be dissolved in the body of another and to encounter its interior *as interiority,* would have a profound impact not only on the evolution of psychoanalysis but also on the nature of representation, especially that of the human body. The X-ray image, its fluid inside-outside perspectives, would, like psychoanalysis, participate in the dissolution of a structure of subjectivity that had sustained Western art and science through the age of the Enlightenment.

• • •

The Enlightenment project, which Ernst Cassirer characterizes as an epistemological movement acutely aware of and fascinated by the contours of the limit, reached a crucial threshold on 8 November 1895, when Wilhelm Conrad Röntgen discovered the X-ray.[21] Enlightenment reason mapped a psychogeography of limits, charting an economy of visuality and a subject seeking to see, expose, and appropriate—according to a presumed power of the gaze—the world around it. The "Enlightenment," write Max Horkheimer and Theodor Adorno, "is totalitarian."[22] Its ethos, what Horkheimer and Adorno refer to as "the mastery of nature," requires a seeing subject that stands outside the limit and frames the field of vision. An Apollonian view of the entire world from outside. From another world. Totality is defined by the limit that divides interiority from exteriority, achieved from without. The persistence of the limit, of the visible world, maintains the viability of such a subject, defined in its encounter with the limit of visuality as such. With the appearance of the X-ray, the subject was forced to concede the limits of the body. Erasing one limit against which it claimed to be outside, the X-ray image, with its simultaneous view of the inside and outside, turned the vantage point of the spectator-subject inside out. The point of view established by the X-ray image is both inside and out. Everything flat, interiority and exteriority rendered equally superficial, the liminal force of the surface has collapsed. Regarding the capacity of the surface to establish meaning in the world, Gilles Deleuze invokes Antonin Artaud's surfaceless body and says: "In this collapse of the surface, the entire world loses its meaning."[23]

Against the field of X-ray vision, the Enlightenment subject lost its vantage point from the outside: the spectral subject now appeared inside the frame, to the extent that a frame remained at all, an aspect of the spectacle. In the X-ray image, everything—which is to say nothing—is visible. Describing Artaud's body as a "body-sieve . . . no longer anything but depth," Deleuze invokes an inherent schizophrenia, one that might apply to the Enlightenment subject in the wake of the X-ray. Deleuze writes:

As there is no surface, the inside and the outside, the container and the contained, no longer have a precise limit; they *plunge into a universal depth* or turn the circle of a present which gets to be more contracted as it is filled. Hence the schizophrenic manner of living the contradiction: either in the deep fissure which traverses the body, or in fragmented parts which encase one another and spin about.[24]

The erasure of the surface (which paradoxically renders the world and its depths and interiorities superficial), the disappearance of a discernible interiority, plunges the subject into a "universal depth." A total and irresistible depth, everywhere. The world is no longer only outside, but also within, inside and out. It moves within and without you. Catherine Waldby says of the X-ray:

> The surface of the body, its demarcation from the world, is dissolved and lost in the image, leaving only the faintest trace, while the relation between depth and surface is reversed. Skeletal structures, conventionally thought of as located at the most recessive depth of the body, appear in co-registration with the body's surface in the x-ray image. Hence skeletal structures are externalized in a double sense: the distinction between inside and outside is suspended in the image, and the trace of the interior is manifest in the exteriority of the radiograph, the artefact itself.[25]

In the X-ray image, the body and the world that surrounds it are lost. No longer inside nor out, within nor without, body and world form a heterogeneous one. (A one that is not one but together, side-by-side, a series of contiguous planes and surfaces, plateaus.) You are in the world, the world is in you. The X-ray can be seen as an image of you and the world, an image forged in the collapse of the surface that separates the two.

The crisis of X-ray visuality struck at the heart of the seeing subject, but also at the very conditions of the visual as such. According to Linda Dalrymple Henderson, Röntgen's discovery of the invisible rays "clearly established the inadequacy of human sense perception and raised fundamental questions about the nature of matter itself."[26] The X-ray forced a collapse of the Enlightenment figure. The metaphors of vision that constituted the Enlightenment were thrust into a literal semiology, which is to say, no semiology but a destruction of the semiological order itself. The absolute radiance unleashed by the X-ray absorbed the subject, enveloping it in a searing light. The Enlightenment subject had become the focus of its own penetrating look, susceptible to the "self-destructive" force that Horkheimer and Adorno identified as Enlightenment practice. One year before the atomic irradiation of Hiroshima and Nagasaki, they warned: "The fully enlightened earth *radiates* disaster triumphant."[27]

Like Freud, who discovered the "secret of dreams" in the dark, Röntgen discovered the new ray in a dark room, in a camera obscura of sorts. "At the time," says Richard Mould, "Röntgen was investigating the phenomena caused by the passage of an electrical discharge from an induction coil through a partially evacuated glass tube. The tube was covered with black paper and the whole room was in complete darkness, yet he observed that, elsewhere in the room, a paper screen covered with the fluorescent material barium platinocyanide became illuminated."[28] A mysterious form of radiation passed through solid objects, casting fluorescent light upon distant surfaces. A light that arrived elsewhere, displaced from its trajectory. "It did not take him long," Mould continues, "to discover that not only black paper, but other objects such as a wooden plank, a thick book and metal sheets, were also penetrated by these X-rays." And flesh. Like a dream, this form of light moved through objects, erased boundaries between solid objects, crossing their internal and external borders. Like a dream, avisual. By fixing the penetrating fluorescence on a photographic plate, Röntgen both discovered and recorded a new type of ray that penetrated organic and inorganic matter and left a shadow of that object on the plate. Simultaneously a new type of ray and a new type of photography. The invisible electromagnetic ray, it would be learned later, consisted of a shorter wavelength (and thus a higher frequency) than visible light, which allowed it to penetrate and illuminate solid matter. Röntgen named the as yet unidentified rays "x." Unknown, secret, illicit.

From the beginning, Röntgen linked the X-ray to photography by fixing his discovery on photographic surfaces. Although Röntgen protested the association of the X-ray with photographic media, claiming that the use of photography had only been "the means to the end," the fusion, or confusion, had already taken hold of the public imagination. X-ray images came to be seen as photographic documents, indelibly marked by their relationship to the superficiality of photographs. They were images of a three-dimensional flatness. One of Röntgen's first published X-ray photographs is of his wife's left hand, which was taken in the final months of 1895, fifty years after his birth and fifty years before the atomic explosions in Japan. The image depicts Berthe's skeletal structure and the bones that constitute her hand, but also the wedding ring that hovers on the surface, infiltrating her hand from the outside. The trace of exteriority that Berthe's ring imposes on the interior dimension reveals the uncanny nature of the new medium. From Irma to Berthe, images of women's interiority appear to have increased after 1895. "The frequently published image of a woman's hand," writes Lisa Cartwright, "gained enormous popularity, becoming an icon of female sexuality and death."[29]

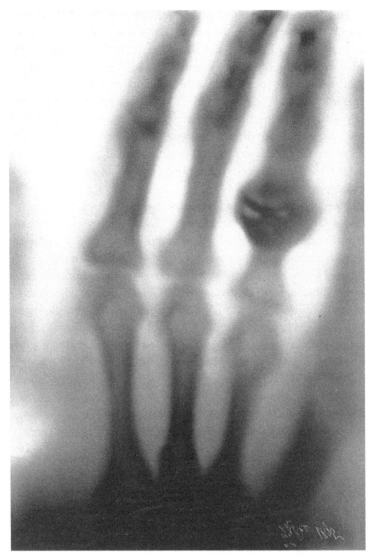

Wilhelm Conrad Röntgen, Berthe Röntgen's X-rayed hand, 1895. Her wedding band is
visible as an opaque mass in the image. Courtesy of Deutsches Museum, Munich.

Cartwright explains, "In the public sphere as in medicine, the female
hand X ray became a fetish object par excellence."[30] "Besides the many
physicians who immediately repeated Roentgen's experiments," Cartwright
says, "a woman's hand, sometimes captioned as 'a lady's hand,' or a 'living
hand,' became a popular test object."[31] But the fascination with X-ray imagery exceeded the medical realms and became a fashion, which Cartwright

describes: "The historian Stanley Reiser relates that 'New York women of fashion had X-rays taken of their hands covered with jewelry, to illustrate that beauty is of the bone and not altogether of the flesh' (or to use a more familiar turn of phrase, is not just skin deep), while married women gave X rays of their hands (presumably with wedding ring affixed, like Berthe Roentgen's hand) to their relatives."[32] The collapse of interiority and exteriority is marked in these images by the convergence of bones and jewelry, whose opacity renders them indistinguishable in the X-ray photograph. One wears one's skeletal structure, the architecture of the body, on the outside. Or rather, interiority and exteriority take place together on the surface. A depth rendered superficial.

On seeing her flesh transgressed, her interiority brought to the surface, Berthe is said to have shuddered at the "vague premonition of death" it evoked.[33] Visible in this image is "the true story of your death": an image of the future, a photography yet to come, your death. Berthe's vision emerges in the encounter with her body, its interiority, which returns to her as if from the outside, from the future. Like Borges's archive and Freud's discovery of formlessness inside Irma's body, Berthe may have glimpsed the avisuality of the fully illuminated self. In this moment of profound intimacy, Berthe peers into the depths of her own body and sees the future, sees her own death, an image of her absence at the center of her body. She is exposed. Interiority and futurity. Berthe recognizes the lethal anniversal force of the photograph. This architecture of the hand, of her hand, comes to Berthe as an alien touch from the irreducible distance of a secret body she carries within her. Of Freud's dream image of Irma, Lacan says:

> The phenomenology of the dream of Irma's injection . . . leads to the apparition of the terrifying anxiety-provoking image, to this real Medusa's head, to the revelation of this something which properly speaking is unnameable, the back of the throat, the complex, unlocatable form, which also makes it into the primitive object *par excellence,* the abyss of the feminine organ from which all life emerges, this gulf of the mouth, in which everything is swallowed up, and no less the image of death in which everything comes to an end.[34]

Medusa's head or acephalic hand. The woman's body—Irma's and Berthe's—comes to determine an archive in which your death is inscribed, formless and secret. In the encounter with radical exteriority, the looking subject disappears. Henderson makes explicit the line between X-rays and the iconography of death, claiming that Röntgen's discovery "triggered the most immediate and widespread reaction to any scientific discovery before the explosion of the first atomic bomb in 1945."[35] "The discovery of x-rays," Henderson concludes, "produced a sense that the world had changed irrevocably."[36]

One hundred years after the discovery of X-rays, the desire for total visibility returned in the form of the Visible Human Project (VHP) and its first embodied figures, the so-called Visible Human Man (Joseph Jernigan, convicted felon, 39, was executed in 1993) and Visible Human Woman (unnamed, "a Maryland housewife," 59, who died of a heart attack in 1995). Immediately after his execution, Jernigan's body was frozen, scanned, and then dissected into one-millimeter planes. Each plane was photographed and digitized over the course of nine months, after which a complete archive of the human body was born. "In this way," Waldby says, "the corpse was converted into a visual archive, a digital copy in the form of a series of planar images."[37] Human bodies transformed into planes of visibility, into thousands of unique visible surfaces, stored in a massive and virtual archive. An archive of the body, of the human body made totally visible in countless possible shapes and configurations. A paradoxical archive, according to Cartwright, at once universal and specific: "The projects share the paradox of seeking to create a universal archive through which to represent and to know human biology, while rendering their respective body models with a level of specificity that may ultimately confound goals such as the establishment of a norm."[38] Waldby likens the advent of the VHP to an inversion of H. G. Wells's "invisible man": "Every structure and organ in the interior of Jernigan's body was about to become an object of exhaustive and globally available visibility."[39] Global visibility: a universal archive, in which everything in the world is visible, and everything is visible to the world.

Even with 3-D and virtual reality techniques, one of the early challenges of the VHP involved reconciling the depth of the body, its volume, with the flatness of the image. It is a problem, according to Waldby, that exists already in the practice of anatomical representation. "The central problem of anatomy is the incommensurability between the opaque volume of the body and the flat, clean surface of the page."[40] The solution, says Waldby, lies in the analogical link between the body and the world, between their figures and modes, between anatomy and cartography. In the case of anatomy, Waldby writes:

> This problem was resolved to some extent through the creation of analogies between anatomical and cartographic space, analogies evident in the fact that the book of anatomy is known as an atlas. If the interior of the body could be thought of and treated as *space,* rather than as a self-enclosed and continuous solid volume, then it could be laid out in ways which are amenable to mapping.[41]

The body shares its figures and modes of spatial representation with the world. In anatomical and geographic maps, both are rendered "as an accretion of laminar 'surfaces,' as landscapes to be traversed by the eye, a volume composed of layers and systems of tissue which are laid one upon the other."[42] The world is a body, the body a world, both *exscribed* in flat space.

From anatomy to the VHP via the X-Ray, the problem of visual representation, of the visibility and visuality of the body, remains located on the surface, on the screen. The body and world, actual and virtual space are *exscribed* on a document, photographic surface, or screen. The VHP "enacts," says Waldby, "the proposition that the interface between virtual and actual space, the screen itself, is permeable, rather than an hygienic and absolute division."[43] A tissue, inside and out, inside out. Waldby traces one genealogy of the VHP to the X-ray, which transgressed the interior of the body, the very structure of the interior as such. She writes:

> The x-ray introduced a form of light which no longer glanced off an inner surface to make it accessible for medical vision but rather cut through the very distinction between inner and outer. Its spectral images rendered the body's interior as irradiable *space* and illuminated scene. . . . The light of the x-ray does not simply penetrate its object, it also projects it, moving through it until its force is interrupted by a screen. Hence the trace of anatomic structure can be both externalized and fixed as radiographs.[44]

The X-ray projects the body's interiority outward until it reaches a screen. The anatomic light of the X-ray *exscribes* the body, irradiates and projects it on an exterior surface. Its destination always a screen, the X-ray achieves a form of completion in the VHP. "The Visible Human figures," says Waldby, abolish "all distinction between surface and depth, demonstrating that all interior spaces are equally superficial, that all depth is only latent surface."[45]

At work from the X-ray to the VHP is also a form of destructive visuality, a visibility born from annihilation. The process of preparing the human body for the VHP archive also annihilated it. The method of dissecting the human body into minute planes "effectively obliterated the body's mass, each planed section dissolving into sawdust due to its extreme desiccation," says Waldby.[46] Like the X-ray, total visibility brought destruction, which is perhaps its condition of possibility. The body reduced to sawdust, to ashes, to cinders. An atomic body, avisual.

Röntgen, the X-ray, photographic media, and the atomic weapon circulate in a specular economy, bound—as are all *photographic* events—by the logic of anniversaries. By capturing single moments in time, all photographs suggest future anniversaries. Individual moments become monumental, arbitrary instants are fixed in the chronic time of anniversaries. Photographs constitute archives and archaeologies of the past but also initiate, like Borges's Library, a "minutely detailed history of the future": in the sense that made Roland Barthes shudder, each photograph inscribes "the true story of your death." It waits for you to arrive to the place of your anniversary, your apocalypse, "now, forever, whenever," as William Haver says.[47] Like the apocalypse that Haver describes, the logic of anniversaries, of photographs, is always accidental, imagined, invented, "unpredictable": anniversaries consist of projections that never adhere, as it were, to the present moment, but always to another time that has passed and will come again. Sometime. Of the apocalypse, an *arkhê*anniversary of sorts, of the temporality that marks the apocalyptic event of my death, Haver writes:

> We must think the apocalyptic as an infinite and indifferent *punctuality*. This punctuality would not be the punctuation that would mark either fate or destiny; neither would it be the punctuation that brings a narrative temporality to term. Punctuality would here indicate, precisely, a nondelay that is not presence, the indifference of the time of material singularity, that entropic "historical space" which would be radical atemporality.[48]

"An infinite and indifferent *punctuality*," like the *punctum,* perhaps, the photographic wound (prick) that Barthes imagines.[49] Punctuality, punctuation, and *punctum:* temporality, inscription, and corporeality. The anniversary can be said to cohere in this indifferent and infinite *punctuality:* inside and outside time, always written or yet to be written, always on the body's surface, either on its inside or outside. An event of writing, of inscription, on the body, there, always—at this precise moment—this or that moment (indifferent yet punctual), "now, forever, whenever." Circumscribed and circumspect. Your death, your apocalypse, your anniversary, always singular and indifferent, for you and whomever else.

Anniversaries are forged by suturing one indifferent moment to another, one *extemporality* to another (one improvised moment outside time to another), toward an infinite singularity; they are inexorably allusive, atemporal, and antihistorical. One moment suggests another, by association, through a secret logic of coincidence; each moment casts a shadow over the future. In the case of 1895, the advent of radiographic imagery permeated by allusion scientific and cultural practices, establishing a kind of spectral episteme that revolves around the representation of interiority. Moving forward and backward in history, 1895 haunts the anniversaries that precede and succeed it. The centennial of Röntgen's birth (1845–1923) and the semicentennial of his discovery, Victor Bouillion notes, coincide with the fiftieth anniversary of the advent of cinema and the annihilation of

Hiroshima and Nagasaki.[50] Expanding the cycle of radiation and mnemic phenomena, the first successful daguerreotype of the sun was taken by Hippolyte Fizeau and J.-B. Léon Foucault on 2 April 1845, one week after the birth of Röntgen on 27 March. From solar to atomic radiation, the anniversaries, measured in fifty-year units, converged in 1995, bringing into sharp focus the thanatographic legacies of the twentieth (XX) century. Some of the events that marked the continuing dynamic of this anniversary included the global festivities commemorating the centennial of cinema; the controversy that erupted over the Smithsonian Institution's failed Enola Gay exhibit; and the awarding of the Nobel Prize for physics to two scientists who were said to have developed techniques that exceeded the capacity of the X-ray.[51]

The X-ray situates the spectacle in its context as a living document even when it depicts, actually and phantasmatically, an image of death or the deterioration of the body that leads to death. A living image of death and the deathly image of life are intertwined in the X-ray. Soon after its discovery, the destructive nature of the X-ray became visible on the human bodies that it pierced. As the euphoria of Röntgen's discovery began to settle, a series of symptoms began to appear on the bodies submitted to the X-ray. Sunburns, hair and nail loss, scaling of the skin, nausea, and an array of other pathogenic signs began to expose the X-ray's destructive capacities. To see and to burn. The two functions and effects are fused in the X-ray, which makes the body visible by burning it. The extravisibility of the X-ray is an effect of its inflammatory force. X visuality. It sees by burning and destroying. An extravisuality that *cinefies*. Under the glare of the X-ray, the body moved from a referent to a sign, from a figure to the primary site of inscription.

X-rays had turned the human body itself into a photographic surface, reproducing its function directly on the human skin. The legacy of photography already contained a version of this fantasy with regard to the human body. Félix Nadar recounts Honoré de Balzac's belief that the human essence comprises "a series of specters, infinitely superimposed layers of foliated film or skin [*pellicules*]."[52] In Balzac's fantasy, a layer of skin is removed and captured each time one poses before a camera. The X-ray amplifies the effects of an intrinsic *photophobia* already encrypted in photography.[53] The direct effects of radiation on the human surface would be reenacted exponentially fifty years later when the atomic explosions in Hiroshima and Nagasaki turned those cities, in the instant of a flash, into massive *cameras;* the victims grafted onto the geography by the radiation, *radiographed.*[54]

As sign, the X-ray assails the movements of visual signification: the referent of the X-ray photograph, for example, trembles between *what* is seen

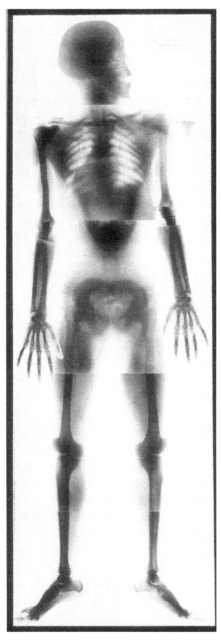

Ludwig Zehnder, an X-ray of the human body, 1896. A montage of nine images. Courtesy of Deutsches Museum, Munich.

and the *process* by which it is seen. For the histories of optics, photography, and phenomenology, the impact of Röntgen's discovery of the X-ray remains immeasurable. Michel Frizot says: "The discovery of X-rays and its effects had considerable repercussions on modern thought. What had been demonstrated was that a completely invisible emanation could manifest its presence on a photographic plate . . . and that this invisible emanation could be used to make the internal reality of the body which was also invisible—a kind of 'inner life' " *(une sorte d'"intériorité").*[55] One hears in this description the resonances of Freud's dream. The convergence of psychology and photography, Frizot suggests, altered the epistemology of the twentieth century, creating, in the process, an epistemology of the inside.

Although it developed primarily within medical and scientific institutions (Röntgen was a professor of physics at the University of Würzburg in Bavaria), the X-ray image has always hovered at the intersection of science and art, technique and fantasy. In the first years after its discovery, X-rays figured in a variety of commercial products and legal considerations: "In London a firm advertised in February 1896 the 'sale of x-ray proof underclothing,' and in the United States Assemblyman Reed of Somerset County, New Jersey, introduced a bill into the state legislature prohibiting the use of X-ray opera glasses in the theaters."[56] The idea of the X-ray, its imagined and imaginary properties, determined the response to its appearance. Still, "others thought," notes Röntgen biographer Otto Glasser, "that with the x-rays base metals could be changed into gold, vivisection outmoded, temperance promoted by showing drunkards the steady deterioration of their systems, and the human soul photographed."[57] The X-ray had become a fantastic repository, an extensive archive of unfulfilled wishes—an unconscious, of sorts. X-rays are "identified in the public mind," says Daniel Tiffany, "with the existence of a world of hidden energies or forms, with visual registrations of invisibility."[58] Many artists have turned to the alluring play of deep surface, or flat depth, in the X-ray: the Futurists, Marcel Duchamp, Frantisek Kupka, Man Ray, and László Moholy-Nagy in the early twentieth century, Jean-Michel Basquiat, Barbara Hammer, Gary Higgins, Alan Montgomery, and Ann Duncan Satterfield, among many others, in the latter half of the XX century. Throughout its history, science and art appear confused in the X-ray, provoking a fundamental problem with regard to its visuality: what does one see *there,* in the X-ray? What constitutes, defines, determines the *thereness* of the X-ray? What is *there* in the X-ray, depth or surface, inside or out? What is *there* to be seen? A *thereness,* perhaps, that is avisual: a secret surface between the inside and out, the place where you are, there, secret and invisible. A spectacle of invisibility, shining, shown, avisual. When technological advances facilitate the appearance of previously unknown phenomena, they often take on the semblance of an artwork. The legacy of the atomic bomb, particularly its spectacular form, attests to this effect.

• • •

In 1946, one year after two atomic explosions incinerated Hiroshima and Nagasaki, the Hungarian artist László Moholy-Nagy (1895–1946)—who also shares a birthday with the X-ray—discussed the impact of X-ray technology on the practice of art: "In x-ray photos," he writes, "structure becomes transparency and transparency manifests structure. The x-ray pictures, to which the futurist has consistently referred, are among the outstanding space-time renderings on the static plane. They give simultaneously the inside and outside, the view of an opaque solid, its outline, but also its inner structure."[59] Moholy-Nagy's description of the use of X-ray technologies in art exceeds mere technique. The ability to simultaneously expose the inside and outside of a thing, to retain the object's surface while probing its depths, describes a scientific fantasy as well as imperative. "Roentgen's initial report," says Cartwright, "was received in the popular press as a discovery of a new force in nature, and not of a new technique."[60] A force rather than a technique, a nature rather than a technology.

X-rays retain the contours of their object while rendering its inside, generating an impossible perspective. Figure and fact, an object's exterior and interior dimensions, are superimposed in the X-ray, simultaneously evoking and complicating the metaphysics of topology in which the exterior signifies deceptive surfaces and appearances while the interior situates truths and essences, what Francis Bacon calls "the brutality of fact."[61] The term *artefact* perhaps best describes the X-ray image, which is at once buried and revealed, invoking its archaeological nature as spectacle. The X-ray image determines a kind of living remnant, a phantom subject. "Art" and "fact," fused together like the sign and referent of Barthes's photograph, arrive as a superimposition in the X-ray, an *artefact*.[62]

Moholy-Nagy posits the fundamental obscurity of X-rays, suggesting a semiology of the image. X-ray images, he says, "have to be studied to reveal their meaning; but once the student has learned their language, he will find them indispensable."[63] A type of exegesis is thus required, a hermeneutics of the X-ray image that, not unlike dream analysis, extends, according to Moholy-Nagy, into the realm of language. A super- or extralanguage, indispensable. X-rays invoke an inherent grammatology. The term *X-ray*, its theatrical prefix, conjures a cultural semiology that includes pornography (X-rated or triple X–rated films), Christianity (Xmas), drugs (ecstasy, or "x"), algebra, signature ("X"), and erasure (crossing out, Malcolm X).

Even before the eruption of radioactive violence in 1945, the signifier "x" was in circulation and overdetermined. Within military circles, Japan's 7 December 1941 strike on Pearl Harbor was code-named "X-day" to signify both the unnameable information embedded within a military communication and the unnameable violence that marked the beginning of the Pacific War. Simultaneously known and unknown, "x" eludes the economy of signification, generating a phantasmatic signifier without signification or, conversely, a full signification with no signifier. "X" can be seen as the master signifier for no signification, for deferred or postponed, over-inscribed and erased signification. An image of imagelessness, a figure for what Tiffany calls the "negativity of the modernist Image."[64] Both a letter of the Roman alphabet, a number, and a figure, a graphic symbol, "x" operates within various economies of signification and meaning at once, never reducible to one system or another, to language or image. A trace: an erasable sign and sign of erasure that erases as it signs and is in turn erased already. On the topographic economy of the trace, Derrida says:

> Since the trace is not a presence but the simulacrum of a presence that dislocates itself, displaces itself, refers itself, it properly has no site—erasure belongs to its structure. And not only the erasure which must always be able to overtake it (without which it would not be a trace but an indestructible and monumental substance), but also the erasure which constitutes it from the outset as a trace, *which situates it as the change of site,* and makes it disappear in its appearance, makes it emerge from itself in its production.[65]

As a residue of the sign, the trace of a sign that erases itself—constituted from the outset as the trace of erasure—"x" signals a "change of site," of the sign, of changing sights, which takes place behind but also on the surface "x." "X" marks or strikes the sign (a sign of the strike, a strike over and against the sign, and the strike itself), replacing the sign as a sign inscribed over the sign, the mark of an illusory or phantasmatic depth. Beneath the "x" opens a phantom depth, inscribed on and as a surface—"x" as the sign for an imaginary topography deep beneath the signifier. "X" signs what is not there, not yet there, or what lies beneath the sign—what was there before the mark "x." "X" marks a spatial and temporal displacement of the sign. A sign that functions like a signifier (like a photograph, according to Barthes's photosemiography). A sign of negative writing: a writing that overinscribes, inscribes itself over inscription, on its surface. "X" obscures what was there, it *xscribes.* An *xscript, xscription, exscription.* A mark on the outside, but also a mark that is no longer a mark, a former mark or *aftermark, ex-scription.*

The rhetoric of photography reveals a constant tension between nature (the signifier of the real) and the photographic referent. From its inception photography was endowed with an organic artifice, seen as a paradoxical form of natural technology. William Henry Fox Talbot describes photography as the "pencil of nature." The critical terminology that frames the photographic practice—"photogenic drawing" (Talbot) or "biogrammatology" (François Dagognet's description of Etienne-Jules Marey's chronophotography), for example—underscores the ambiguity between figure and fact in photography.[66] The X-ray expanded not only the limits of the empirical world and the human sensorium but also the trajectory of photography: it extended the graphic reach of the apparatus into the invisible spectrum of light. It introduced *exscriptive* writing, an inside-out writing that reverses the trajectory of Enlightenment writing from inscription to *exscription*. The mark is no longer made from the outside in nor, for that matter, from inside out: writing takes place outside, it remains irreducibly elsewhere, an *exscriptive* mark that never adheres to the interiority of a text or document—displaced, atopic, and atextual.

• • •

Several coincidences punctuate 1895, bound by a specular and ultimately phantasmatic historicity. During 1895, Freud, the Lumières, and Röntgen made significant advances in their respective fields, initiating a heterogeneous history of visuality that remains illegible, invisible, and ahistorical.[67] No historiography can produce an adequate narrative for the events of 1895, which are themselves secure only as momentary designations for events that, like most historical processes, take place over time. A phantasmatic history, then, imagined and imaginary. A coincidence of histories, a history of coincidences, bound only by a phantom chiasmus, "x."[68]

On 13 February of the year that Röntgen published his discovery of the X-ray, Louis Lumière patented the Cinématographe.[69] The Lumières' first public screenings were held at the Grand Café in Paris on 28 December, the *exact* date of Röntgen's publication.[70] Louis Lumière's idea to adapt the drive device of a sewing machine to the Cinématographe came to him "one night," says Emmanuelle Toulet, citing Lumière: "One night when I was unable to sleep, the solution came clearly to my mind."[71] Like Freud's dream, the secret of cinema, its solution was revealed to Lumière at night. The dream of cinema. The original configuration of the screening space included a translucent screen, hung in the middle of the chamber, which passed the image in reverse to the other side. Spectators could pay more to sit on the side of the projector or less to view the screenings from the other side. The contemporary organization of the theater, which limits the viewing surface to one side, developed later. In this sense, the original apparatus with its transverse flow of light more closely resembled the luminous economy of the X-ray. The first film screen was itself a kind of porous tissue.

One film presented by the Lumières on 28 December, *Leaving the Lumière Factory* (*Sortie d'usine,* no. 91, 1895), remains as an emblem of the early cinema. Shot in front of the Lumière factory, the version shown at the Grand Café was in fact its third remake.[72] The brief *actualité* opens onto a frontal view of the Lumière factory. The factory doors open and workers emerge from the back of the frame toward the front. They move toward the camera, before veering to either side. Two large doors toward the right of the frame open inward, and workers begin to stream outward from it and from a smaller door on the left of the frame. They emerge at different speeds, initially aiming toward the camera before breaking to the right or left. Among the women and men are a dog and several bicyclists. As the factory empties, and before the end of the film, one man rushes against the flow back into the nearly empty factory, reversing the outward, background-to-foreground trajectory. In the distance, a solitary figure remains visible inside the factory, in the distance, deep inside.

The very architecture of the film, the series of flat surfaces that moves from screen to wall to interior background, suggests that this cinema is an exploration of depth. An imaginary, impossible depth that extends into the screen, that opens behind it, revealing a virtual interiority and distance, far away. Noël Burch describes the sensation of depth this film produces:

> Although a wall occupies half the picture, the sense of space and depth
> which was to strike all the early spectators of Lumière's films is already
> present in the contrast between this wall blocking the background to the
> left and the movement of the crowd emerging from the dark interior on
> the right.[73]

This "sense of space and depth" is a general feature of all the Lumière films, establishing an axis along which objects (people, animals, and other animate things) move toward and away from the spectator. An extra space, an extra dimension that exists only as an effect of cinema projection. An abyss, *mise-en-abîme,* abyssal space. The deep space opens only there, in an avisual world, folded from the outside in and the inside out. It has no reference, indexical or otherwise, to any place outside the film, although, as a photograph, it originated somewhere. A space that is not really there. An archive where there is no space. Like the unconscious, like the X-rayed body, an abyss.

In May 1895 Josef Breuer and Sigmund Freud published *Studies on Hysteria,* after years of treating hysterics with hypnosis.[74] The *birth* of psychoanalysis in 1895 was followed by at least two significant afterbirths: the "dream of Irma's injection" in July and the birth of Anna, the last of Freud's children, in December. Using a term shared by film and psychoanalysis, Freud wrote in 1915, "A dream is, among other things, a *projection:* an exter-

Auguste and Louis Lumière, *Leaving the Lumière Factory* (*Sortie d'usine*, no. 91, 1895).

nalization of an internal process."[75] To further overdetermine the shape of 1895—overdetermination being, among other things, a key feature of Freud's typology—Thomas Edison, according to Henderson, asserted confidently "that x rays would ultimately unveil the activity of the human brain."[76] Cartwright and Brian Goldfarb document Edison's and Louis Lumière's immediate efforts to explore X-ray imaging of the body and brain.[77] Brain, body, mind, and soul appear to orbit 1895 in a specular constellation that blurs the distinctions between each form of interiority. The material and immaterial, anatomical and spiritual, psychical and organic aspects of human interiority are confused by the rapid proliferation of imagining techniques and devices in 1895. Psychoanalysis, X-ray, cinema offer new possibilities for the organization of interiority, new designs, new organs. Moholy-Nagy describes the sense of confusion that erupted between scientific developments and desire in the late nineteenth century:

> In the 19th century telescopic and microscopic "miracles," x-ray and infrared penetrations were substituted for fantasy and emotional longing. These phenomena, motion and speed, electricity and wireless, seemed to give food enough to the imagination without introducing subconscious automatism. Photography was the golden key opening the door to the wonders of the external universe to everyone. The astonishing records of this period were *objective* representations, though they went in some cases beyond the observation capacity of our eyes as in the high speed, micro-macro, x-ray, infrared and similar types of photography. This was the period of "realism" in photography.[78]

Inside and outside are lost in the fantasies of realism and desire. For Moholy-Nagy, photography represents the visualization of desire in the nineteenth century, a realist exteriorization of "fantasy and emotional longing." It locates the topology of the unconscious prior to the advent of psychoanalysis, serving as the sign for an unconscious yet to be named.

Three *phenomenologies of the inside* haunt 1895: psychoanalysis, X-ray, and cinema seek to expose, respectively, the depths of the psyche, body, and movements of life. These three technologies introduced new signifiers of interiority, which changed the terms by which interiority was conceived, imagined, and viewed. They transformed the structure of visual perception, shifting the terms of vision from phenomenal to phantasmatic registers, from a perceived visuality to an imagined one. From visual to avisual.

Psychoanalysis, X-ray, and cinema appear fused to one another in a het-erogenealogy of the inside — each seems to appropriate another's features, functions, and rhetorical modes. The capacity to see through the surface of the object, to penetrate its screen, emerged in 1895 as the unconscious of the Enlightenment, but also as an unimaginable light. Benjamin recognized the proximity of photography and film to psychoanalysis, as did Jean-Martin Charcot, Albert Londe, and Freud before him, finding in these two probes two passages toward the discovery of the unconscious. "Evidently," writes Benjamin in "The Work of Art in the Age of Mechanical Reproduction," in 1935–36, "a different nature opens itself to the camera than to the naked eye — if only because an *unconsciously penetrated space* is substituted for a space consciously explored by man. . . . The camera introduces us to unconscious optics as does psychoanalysis to unconscious impulses."[79] Another nature, different and displaced, emerges from an unconscious optics. In Benjamin's language, the movement from photography to the unconscious develops from the logic of penetration.[80] A penetration, like the one that pierces the surface of Irma's dream body, leads to a phantasmatic architectonics of the psyche and the secret body it *exscribes*.

Psychoanalysis, X-ray, and cinema introduce complex signs of interiority that resemble antisigns or traces of signs that signal the view of an impossible interiority. Each sign functions more as a trace than a signifier, each linked with the force of an irreducible exteriority. Derrida says of the trace, of *arkhê*writing: "This trace is the opening of the first exteriority in general, the enigmatic relationship of the living to its other and of an inside to an outside: spacing."[81] A sign of the outside. An ex-sign. Psychoanalysis, X-ray, and cinema: trace of exteriority and interior *design*. A design, both as an architectural imprint or sketch and as an ex-sign, a designation of interiority. A design of the inside, a designation of its contours, but also its appearance as a form of designification. An exemplary design.

3. Cinema Surface Design

Writing in 1928, the impressionist filmmaker Germaine Dulac called for a return to "visual" cinema, a restoration of cinema to its origins in the world of visuality. Narrative form—the structural demands of temporality, continuity, and causality—had driven cinema further from its essence as a visual art. "The great pity, as far as film is concerned," she says, "is that, though a uniquely visual art, it does not at present seek its emotion in the pure optic sense."[1] Cinema operates according to modes of visuality, says Dulac, striving toward a "pure optic sense." Drama and affect in cinema, she insists, "*must* be visual and not literary," emerging from "optical harmonies."[2] In Dulac's idiom, neither "optical harmony" nor "pure optic sense" restricts cinema to the register of vision. Dulac sees in cinema optics a form of tactility and imagines film spectatorship as physical contact. Technical advances allow images to "caress" the eye, sending to it "radiations which touch it more powerfully."[3] Visuality encompasses the senses; pure optics exceed the visible world.

For Maurice Merleau-Ponty, the conception of "light as an action by contact" can be traced to René Descartes. "The blind, says Descartes, 'see with their hands.' The Cartesian concept of vision is modeled after the sense of touch."[4] Laura U. Marks argues that cinema induces a haptic perception, establishing in the registers of the visible a tactile materiality. Following Noël Burch's and Gilles Deleuze's use of the term *haptic* to describe certain modes of tactile representation in film, Marks says: "Haptic looking tends to rest on the surface of its object rather than to plunge into depth, not to distinguish form so much as to discern texture."[5] A mode of visuality that displaces opticality, rendering the visible avisual. Like the secret visibility or absolute invisibility that both Merleau-Ponty and Jacques Derrida describe, Dulac's visuality comes to resemble something other than a conventional economy of vision. It suggests a form of penetrating visuality that deflects the look away from the register of vision and returns it to the subject as another sense. It transforms the field of visuality into a broader sensual order. It effects an embodied visuality, in Marks's sense, but also a psychological visuality. Surface and depth, body and psyche dictate the dual registers of Dulac's visuality.

"Cinema," says Dulac, "by decomposing movement, *makes us see,* analytically…the psychology of movement."[6] Dulac's idiom, which synthesizes metaphors of corporeality, psychology, and perception, reflects the profound dilemma that cinema poses with regard to the rhetoric of visuality. The penetrating visuality of film pierces the surface and exposes "the psychology of movement," the interiority of things. For Dulac, cinema offers the possibility of decomposing the movement of things and revealing, in the process, its psychology, or what she calls elsewhere its "soul."[7] The medium's purpose, she says, is to facilitate a phenomenology of the imperceptible. "If machines decompose movement and set out to explore the realm of the infinitely small in nature, it is in order to visually reveal to us the beauties and charms that our eye, a feeble lens, does not perceive."[8] As the camera eye penetrates deeper, the world decomposes under the glare of its probe: the visuality of cinema renders the world formless, decomposed, reduced to the raw psychology of movements. Walter Benjamin regards the development of "close-up" and "slow motion" effects in film as a way to not only enhance perception but conceive new phenomena. He writes:

> With the close-up, space expands; with slow motion, movement is extended. The enlargement of a snapshot does not simply render more precise what in any case was visible, though unclear: it reveals entirely new structural formations of the subject. So, too, slow motion not only presents familiar qualities of movement but reveals in them entirely unknown ones.[9]

Inside familiar movements, Benjamin concludes, the cinema reveals new and unknown movements, new subjects of motion. He suggests a psychology or interiority of movement, a spatial dimension established by movement, but also within it. Space expands and contracts in the cinema, movement extends and withdraws. Space and movement are malleable and independent of their photographic referents: the spatial and temporal dimensions of each film are unique to that film. Each film produces its own spatiality and temporality. On the screen and on its other side, but also *inside.*

From the first moments of cinema, early practitioners imagined the possibility of revealing a visibility unique to film. In their *History of the Kinetograph, Kinetoscope, and Kinetophonograph,* published in 1895, W. K. L. Dickson and Antonia Dickson imagine the possibilities of a technological simulacrum that makes the invisible visible, the familiar and unseen horrific. They describe their early experiments with kinetographies of "the infinitesimal," films of insects. "A series of inch-large shapes then springs into view, magnified stereoptically to nearly three feet each, gruesome beyond power of expression, and exhibiting an indescribable celerity and rage."[10] That which is invisible in daily life erupts in a nightmarish display of monstrosity, amplified by "celerity and rage." From the depths of the invisible and infinitesimal, life becomes affected with horror and rage. "Monsters close upon each other in a blind and indiscriminate attack," they continue, "limbs are dismembered, gory globules are tapped, whole battalions disappear from view."[11] Stereoptic and magnified, the microscopic world of the Dicksons takes on a fantastic existence. It erupts from the depths of invisibility onto the screen.

By making visible the ordinarily invisible, imperceptible, microscopic life of the world, the kinetograph transforms the movements of life into emotional phenomena. Enhanced visuality, according to the heterogeneous logic that fuses visibility with affect, produces an emotional visuality, an affective view of the world. Brought into view, these insects become something other, charged with unbridled rage and aggression. The Dicksons' description raises questions about the very visibility of the spectacle. "A curious feature of the performance," they add, "is the passing of these creatures in and out of focus, appearing sometimes as huge and distorted shadows, then springing into the reality of their own size and proportions."[12] "The cinema makes us spectators of its bursts of light and air, by capturing its unconscious, instinctive and mechanical movements," says Dulac.[13] Cinema generates a spectacle of the unconscious, of the unconscious of light, rendering its viewers unconscious spectators.

Among the first set of films exhibited by Auguste and Louis Lumière is *The Arrival of a Train at La Ciotat (Arrivée d'un train en gare de la Ciotat,*

Auguste and Louis Lumière, *The Arrival of a Train at La Ciotat* (*Arrivée d'un train en gare de la Ciotat*, no. 653, 1895).

no. 653, 1895). A view from the platform reveals the tracks and the background, which recedes to the right of the screen, into the distance, in anticipation of the train. After a few seconds, the train appears, emerging from the background and proceeding toward the foreground from right to left. The horizontal movement of some passengers is superseded by the force of a diagonal verticality—of the train and departing passengers—or, rather, by the acute sense of deep space the film engenders. This space is supplemental, folded into the screen like a psyche.

The film is known for the apparent shock it induced—as Tom Gunning says, an effect of the train colliding with the film's surface, and thus with the space of the spectator. The emphasis on the illusion of reality, the spectator's fear of the real train, Gunning insists, has been exaggerated. According to those accounts, he says, "Credulity overwhelms all else, the physical reflex signaling a visual trauma. Thus conceived, the myth of initial terror defines film's power as its unprecedented realism, its ability to convince spectators that the moving image was, in fact, palpable and dangerous, bearing towards them with physical impact. The image had taken life, swallowing, in its relentless force, any consideration of representation— the imaginary perceived as real."[14] Although Gunning raises doubts about the veracity of the accounts of early Lumière screenings, he sees in that moment of primitive cinema the capacity to produce a profound sense of astonishment and shock. In Gunning's rhetoric, the image-become-life— and not the train—threatens to breach the space of the spectator and swallow him, her. Fear is displaced from the train to the image itself. "What is displayed before the audience," he says, "is less the impending speed of the train than the force of the cinematic apparatus."[15] The spectator at this screening is "overwhelmed" and "swallowed" by the image, of which the train is its only figure.[16] The description of spectator shock, whether as genuine fear and trauma or as a more subdued form of astonishment, suggests a complex division of space between film and spectator at the surface of the screen.

The spectator is swallowed by the image, as if it were an oral cavity, as if the image, in this instant, revealed an interiority, vast and terrible. An interiority of cinema, a psychology entered like Irma through the mouth. "We cannot," Gunning concludes, "simply swallow whole the image of the naïve spectator, whose reaction to the image is one of simple belief and panic; it needs digesting."[17] The screening and its accounts are marked, in Gunning's idiom, by an economy of digestion, a geology of threatening interiority. The imagined or impending collision at the site of the screen between the train and the spectator can be figured as an ingestion: what awaits the spectator at the projected point of collision is an imaginary depth, a volume that opens onto the spectator from the other side of the screen. A gaping orifice. Gunning quotes one journalist's response to an early screening of a Biograph train film: "An unseen energy swallowing space."[18] The film is an energy, cinema a space (with, it seems, an orifice), the gesture that brings them into contact with the spectator, swallowing. The complex interaction of the two fuses these elements, unseen. The unseen energy swallows, but also renders space, makes space possible and visible, what Deleuze calls "a force."[19] Gunning invokes Michael Fried's categories of absorption and theatricality, positioning early cinema on the order of theatricality. "These early films explicitly acknowledge their spectator, *seeming to reach outward and confront*. Contemplative absorption is impossible here.

The viewer's curiosity is aroused and fulfilled through a marked encounter, a direct stimulus, a succession of shocks."[20] Theatricality rather than absorption, shock rather than contemplation, "a cinema of instants rather than developing situations."[21] But in a sense these films can be seen as absorptive *and* theatrical, following Fried's definitions: the absorptive nature of the image an effect of theatricality, the theatricality an effect of the absorption. One folded inside the other, "the image's absorption in itself," says Fried.[22] Or, the theatrical presentation of absorption. What is presented to the spectator is a theater of absorption, not only in the figure of the train, which threatens to penetrate the spectator's space, but in the projection of "an unseen energy swallowing space."

Other displaced collisions, sites of contact between the spaces of cinema and spectator, can be found in a frequent motif of those early films, the car accident. Automobile explosions and collisions occupy a significant amount of early screen space. Cecil Hepworth's *How It Feels to Be Run Over* (1900), for example, features an automobile plunging directly into the camera, blackening the screen's surface and the viewer's visual field.[23] The actualization of the Lumière train threat begins with a feint. The film opens with a slightly low angle shot of a dirt road, not exactly centered, which stretches into the background. From the first shot, a vehicle is visible in the distance. It moves toward the camera, but at an oblique angle. Gradually, the vehicle becomes more clearly visible as a horse-drawn carriage. The trajectory of its movement suggests it will pass by the camera, which it does, to the right of the frame. Before the carriage has disappeared, a second vehicle appears in the distance. It is moving at a faster pace and directly toward the camera. Through a cloud of dust raised by the passage of the carriage, the second vehicle approaches the camera, clearly visible now as an automobile moving rapidly. Three passengers signal to the camera, to an unseen figure, an extension of the spectator, to move away. The automobile wavers, then from a position slightly to the right of the frame, it crashes into the camera. At the point of impact, a concave bumper *swallows* the space into a black screen, where a phrase appears in separate frames: "? !!? . . . !!! . . . ! . . . Oh! . . . Mother . . . *will* . . . be . . . pleased."

R. W. Paul's 1906 film *The Motorist* (directed by Walter R. Booth) opens with a street scene, an automobile moving toward the camera. A policeman steps into the vehicle's trajectory from the right side of the frame, interrupting the anticipated crash into the camera. The vehicle continues into the policeman, lifting him onto the car's hood and carrying him with it. The automobile passes the camera to its left and exits the frame at the lower-left corner. After a cut, the automobile reenters the frame on the upper-left corner, moving to the right. The policeman, still on the hood of the car (and replaced in this shot by a dummy), is thrown to the ground and run over. As the wheels pass over the figure of the policeman, an immediate cut replaces the dummy with a living actor, who rises from his assault angry but uninjured. The policeman chases the motorists, and the remainder of the film consists of an extended pursuit, which involves, at one point, an ascent into outer space.

In *The Motorist,* the anticipated contact between the automobile and camera/spectator has been inscribed within the diegesis. The figure of the policeman absorbs the impact and defies the laws of nature by returning to life unharmed by the force of the automobile. Inside and out, this side and the other, before and behind have been displaced to the site of a uniquely filmic metaphysics: the fluid boundary between life and death. Paul's *Extraordinary Cab Accident* (1903) deploys the same Méliès cut, the substitution of inanimate objects for living beings in order to display unflinching views of death. A man on the sidewalk kisses a woman good-bye, then steps backward into the street and in front of an oncoming horse and carriage, which knocks him to the ground, then gruesomely runs him over. A policeman who sees the accident chases the carriage, exiting the frame to the lower right. The woman rushes to the fallen man, followed by another man who enters the frame from the lower-left area. They examine the body and declare the wounded man dead. As they acknowledge the man's death, the policeman returns to the crime scene with the perpetrator, and all four pay their respects. Suddenly, the dead man springs up, pushes the policeman, and, grabbing the woman's hand, rushes out of the frame to the lower right. The film is made possible through several continuity cuts that substitute the living body with a dummy, then restore the living body. The movement between the animate and inanimate is achieved by crossing an invisible threshold between the living and nonliving, spaces rendered contiguous, fluid, and reversible in cinema.[24]

James Williamson's 1905 film *An Interesting Story,* about a man so engrossed in his book that he fails to notice the often dangerous world about him, contains a scene in which the distracted reader walks directly into an oncoming steamroller. Using stop-action cinematography, Williamson substitutes a flexible figure for the human actor and flattens it under the steamroller. The flattened protagonist becomes a figure for what Vivian Sobchack calls the "deflation of space."[25] Two men on bicycles rush to the scene of the flattened reader and, using their tire pumps, reinflate him. The reanimated reader thanks his rescuers, then continues forward, into the distance. In each example, a fatal encounter between a human being and a vehicle is reversed; the human being survives, impossibly, and returns to life after having once crossed over to the other side. An orphic cinema of encounter displaced from the "primal scene" of cinema, the encounter between the spectator and the apparatus, its unseen energy, into the diegesis or representation. A *mise-en-abîme* of encounters.

The Lumières' *Arrival of a Train* anticipates the early cinema of encounter, the excitement of a confrontation with the unseen energy of cinema space. Edwin S. Porter's so-called floating shot, a medium shot of a bandit firing a gun directly into the camera (screened sometimes before and at other times after Porter's 1903 *The Great Train Robbery*) fulfills, perhaps, the fantasy of an encounter with the other side, an ultimately phantasmatic encounter as well as other side.[26] *Arrival of a Train* serves, in its brief scope, as a calculation of deep space, a measurement of depth on the surface of the screen. The train, described by some as a metonymy of the cinematograph, connects the film's background and foreground spaces. And like the workers who emerge from the factory in *Leaving the Lumière Factory,* the passengers in *Arrival of a Train* step out of the cavern of the train, another space embedded within the film's visible space. An extra space folded onto the screen, which makes possible yet another plane, another dimension that extends the fantastic depth of cinema. In this film, the train can be said to transport not only its passengers, to suture the distant and close spaces of the screen, but also space: a private, discrete, and portable space that opens onto the screen, revealing its interior cargo, namely, passengers.[27]

Arrival of a Train stages a collision between the image and the spectator through the vehicle of a vehicle, a figure of the figure of encounter, the train. The spectator and train never collide, but the spectator and image do. The spectator and moving image meet, in *Arrival of a Train,* at a particular point in space, the screen, which becomes a phantasmal and *displaced* site, what Deleuze calls a "metaphysical surface." Following the Stoics and their thinking of a physics of the surface, in which the world is organized and made *sensible* on the surface, Deleuze says:

> There is therefore an entire physics of surfaces as the effect of deep mixtures—a physics which endlessly assembles the variations and pulsations of the entire universe, enveloping them inside these mobile limits. And to the physics of surfaces a metaphysical surface necessarily corresponds. Metaphysical surface *(transcendental field)* is the name that will be given to the frontier established, on the one hand, between bodies taken together and inside the limits which envelop them, and on the other, propositions in general.[28]

Within, behind, or beneath the surface, a set of "mobile limits," a metaphysical surface opens up. In it, not only new organizations of space, new relations "between bodies taken together and inside the limits which envelop them," but new frontiers, possibilities of "propositions in general." New propositions that regulate the very orders (laws) and interactions of space as such. The cinema screen separates space, establishes orders and relations between phenomenal and existential, if not metaphysical space: the space between life and its shadow, but also between discrete orders of life, movement, and animation.[29] The screen is a deep surface that brings together two velocities in an imminent collision. A point of contact between screen elements but also between "propositions in general." *Arrival of a Train* makes the point of contact (which has to remain ultimately deferred) visible; a point that is an opening, a crack that leads to the other side.

In a countergesture to *Arrival of a Train*, the Lumières' *Leaving Jerusalem by Railway* (1896) depicts a train station in Jerusalem from the perspective of the train. As the train pulls away from the station, a camera mounted on the train's rear films the view of the station and people, some of whom address the camera, move toward it, away from it, and others, indifferent, who stand still as the train moves farther away. *Leaving Jerusalem* uses human figures to mark coordinates in deep space, to chart a virtual space and phantasmatic volume that opens up *inside* the screen. In *Leaving Jerusalem*, the train has merged with the camera moved inside it, as it were, collapsing the figure of the train into the camera and erasing the metaphor by fusing it with the screen's surface.

A 1905 Biograph film, *Interior New York Subway, 14th Street to 42nd Street*, directed by G. W. Bitzer, further extends the reversal of *Leaving Jerusalem*.[30] The approximately five-minute film consists of a series of continuous shots of a subway train, seen from behind, as it travels through the dark interiors of the New York underground in 1905. The camera, mounted on a second vehicle, follows the train as it moves away from the camera and plunges farther into the deep space of the screen.[31] When the first train stops at each station, the second vehicle stops and pauses, the camera still filming, until the train resumes its travel. The lighting fluctuates, depending on the train's location with respect to other trains and stations, and the passage of light through the vertical beams and columns on either side of the train creates a flickering effect, a metonymy perhaps of the projector, a *mise-an-abîme* inscription of it on the screen's surface.[32] The "whitish grey" structures appear skeletal, perhaps in the way that Irma's interiority looked to Freud in his dream, as if the subway represented a vast and abstract corporeality. An archive of the underground and an underground archive. The dark horizon and destination of the train are limitless, infinite, unascertainable. Vaguely formless. The train continues forward, moving farther into deep black space, an interiority of the film that seems to extend endlessly. Eventually the train stops at a station for an extended period of

G. W. Bitzer, *Interior New York Subway, Fourteenth Street to Forty-second Street* (1905).

roughly thirty seconds, while passengers exit. But before the film ends, the train and the camera that pursues it resume their forward movement, suggesting a continuation of the train's progression into the abyss.

From the Lumières to Bitzer, the passage of the train forward and backward, toward and away from the screen and spectator, determines a trope of early cinema, the construction of deep space in two dimensions. Gunning's early cinema of attraction might also include the production of a synthetic volume and imaginary depth that opens up on the screen. Other Lumière films reveal a sensitivity to receding space and planes of depth. A Lumière film that echoes Gunning's idiom of eating and ingestion, *The Baby's Meal* (*Repas de bébé,* no. 88, 1895) can be seen as a series of planes made visible by the distances of leaves rustling in the background. Auguste and his family occupy the center of the frame, in medium shot. The horizontal line of the table, which slices across the screen's lower half, is intersected at an angle by the side of the house, which pulls the space of the film into the background. In the brief scene that features the Lumières eating, various planes of deep space are stirred by the animated foliage that surrounds the family. The activity of leaves extends the space backward into a seemingly vast depth that opens up behind the family and threatens to engulf them. An abyss that swallows the eating family, a *mise-en-abîme* of oral cavities.

In those first moments of cinema, the spectacle unfolds as an energetic space, as an avisuality of spaces that are there but unseen, as the avisuality of unseen energies. Williamson's film *The Big Swallow* (circa 1901) *disfigures* the metaphors of digestion and absorption, rendering the deep spaces and "unseen energies" of cinema as a literal mouth "swallowing space." The three-shot comedy begins with a three-quarter shot of a man who appears to be gesturing angrily toward an unseen camera. He approaches the camera in a theatrical address, gesturing his adamant refusal to be filmed. As he moves from medium shot to close-up, he opens his mouth and continues to move toward the camera. The dark cavity of his mouth fills the screen until it coincides with the frame, engulfing the entire screen. The mouth produces, as it were, an opening to the other side, to the inside; the black screen a view of interiority. The other side inside. In the darkness a cut, then the camera and operator enter the picture and plunge—are swallowed—into the dark screen-mouth. They vanish, and after another cut, the scene recedes and the man reappears, in extreme close-up, smacking his lips.[33] He has swallowed the camera and operator. *The Big Swallow* ends in a close-up shot of the laughing, digesting man.

The film's conclusion introduces a logical inconsistency: if the camera has been swallowed by the subject, then how can the film continue? Who

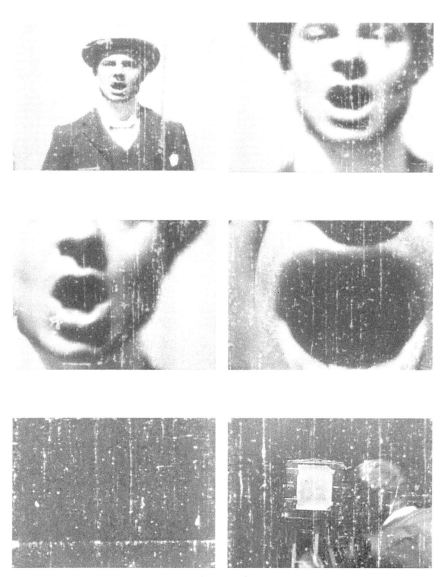

James Williamson, *The Big Swallow* (circa 1901).

remains to finish the film; who or what is there, at the site of the camera? The camera and its double have been introduced into the film, the entire film rendered a phantasm. One remains on this side, the other has crossed over, passed through the screen through an orifice, to the other side, into the other space of the film. Absorbed into the film's world and interior space, the function of the apparatus has moved from the camera to the spectator, who now occupies the camera's place. The spectator is left to secure the vacant but still vigilant function of the apparatus. It has become the camera, its trace, an avisual aftereffect or image. The film's subject has moved from a spectator to a specter, to a spectral phantasm.

The Big Swallow actualizes a trope of early cinema, spatial depth and volume. In the black space of an extreme close-up, another world opens on the surface.[34] In contrast to the frame of a painting, says André Bazin, which binds the space of the image inward, the edges of the film screen indicate a vast expanse that opens outward from the screen. "The picture frame polarizes space inwards. On the contrary, what the screen shows us seems to be part of something prolonged indefinitely *into the universe.* A frame is centripetal, the screen centrifugal."[35] Inside Williamson's black screen, inside and beyond it, an abyss embodied: a body and screen, fused, swallowing space. The movement into the man's mouth echoes, perhaps, Freud's X-ray entry into Irma's mouth. It marks the passage of the subject into the illusory body of an other, but also the loss of oneself elsewhere, in an other, deep inside an other. "The mouth is not only a superficial oral zone but also the organ of depths," Deleuze says.[36] Eager to see, the camera and its operator have plunged into the depths of a human subject, invoking a kind of film psychology, a deep interiority, an unconscious or figure of the unconscious in the form of an avisual volume represented by a dark screen. "A psychology of movement," perhaps, as Dulac says, or a movement of psychology, a mobile psychology. In place of the camera, in its place, the spectator. You. You are there, in its place, a lost object.[37] A trace of the camera, its phantasm. At the surface then, in the virtual space that sutures and severs inside from out, here from there, a phantasm erupts. In you.

• • •

A phantasmatic geography of the subject stretches across the metaphysical surface of the screen. Metaphysical—like the architecture of Borges's Library, Tanizaki's dark house, Irma's mouth, Berthe's X-rayed hand, and cinema—because this surface establishes an impossible order of deep space. A superfluous and superficial physics. A deep flatness on the surface that allows one to "slide" inside and out, from this side to the other side and back. About sliding, Deleuze says:

By sliding, one passes to the other side, since the other side is nothing
but the opposite direction. If there is nothing to see behind the curtain,
it is because everything is visible, or rather all possible science is along the
length of the curtain. It suffices to follow it far enough, precisely enough,
and superficially enough, in order to reverse sides and to make the right
side become the left or vice versa.[38]

Both sides are connected by the surface, by the tissues of a metaphysical
surface that renders the world and its geography superficial. Neither side is
absolute nor oppositional, merely opposite, connected by some corner or
fold. Inside and outside contiguous, everything visible. Depth, Merleau-
Ponty says, understood as a "dimension that contains all the others is no
longer a dimension":

Depth thus understood is, rather, the experience of the reversibility of
dimensions, of a global "locality"—everything in the same place at the
same time, a locality from which height, width, and depth are abstracted,
of a voluminosity we express in a word when we say a thing is *there*.[39]

A dimension that contains all other dimensions, each one reversible, in
which everything is visible "in the same place at the same time," *there*. A
flat depth, a depth on the surface, in which "height, width, and depth are
abstracted." An abstract, superficial volume. (Like the X-ray image, avisuality may be the effect of total visibility. If "everything is visible," then the
economy of vision is no longer regulated by presence and absence, visibility and invisibility, but rather by the play of phantasms.) The interior and
exterior designed, reversible, indistinguishable in the realm of total visibility. A phantasmatic design of the world, which has lost its coordination, its
signs and singularities. Unsigned and *designed*.

The *designed* world that opens up in the mouth of cinema is unconscious in at least one respect—it shares with Freud's archive the absence of
negation. Inside and out, this side and that other side, life and death, you
and your double, "the true story of your death" and all others are viable in
cinema, in the specular archive that extends in all directions and across all
temporalities without limits. (Timelessness, in Freud's definition, is also a
structural aspect of the unconscious.) A prevalent feature of what Noël
Burch calls "the primitive mode of representation" (PMR) in film consists,
he says, of "non-closure." In contrast to the closures of institutional representation, which seek to center the spectator as the subject of representation, the nonclosure of the PMR radically displaces the subject from the
site of the spectacle. Burch writes:

If institutional closure is taken to be more than narrative self-sufficiency
and a certain way of bringing the narrative to an end, if, on the contrary, it

is treated as the sum of all signifying systems that centre the subject and lay the basis for a full diegetic effect, including even the context of projection, then the primitive cinema is indeed non-closed as a whole.[40]

The nonclosure of early cinema effects a field of representation, a "sum of all signifying systems," which is limitless, like Borges's Library, and formless. It thrusts the subject of the spectacle into an abyss, into and through an orifice that opens phantasmatically on the screen surface. If cinema is unconscious, is a form of or like the unconscious, it is because cinema shares with the unconscious a fantastic mobility and geography; both modes of avisuality regulate, vis-à-vis the metaphysical surface, the passage of a phantasm, from here to there, this side to that.[41] Through an avisual hole. Psychoanalysis, too, says Deleuze, is a superficial practice, a practice and thought deeply attuned to the properties of the surface, in its metaphoric and material forms: "The surface has a decisive importance in the development of the ego, as Freud clearly demonstrated when he said that the perception-consciousness is localized on the membrane formed at the surface of the protoplasmic vesicle."[42] Deleuze refers to Freud's attempt to figure, in *Beyond the Pleasure Principle* (1920), perception-consciousness *(Pcpt.-Cs.)* according to a model from evolutionary biology, requiring the development of a theory of surfaces. Freud begins with an apology and a speculation.

"What follows is a speculation," says Freud, "often far-fetched speculation, which the reader will consider or dismiss according to his individual predilection."[43] Since *Pcpt.-Cs.*, the mechanism by which stimuli from the outside world and feelings of pleasure and unpleasure from the inside are processed into consciousness, must be accessible from the inside and out, "it is therefore possible to assign to the system *Pcpt.-Cs.* a position in space."[44] "It must lie on the borderline between outside and inside; it must be turned towards the external world and must envelop the other psychical systems."[45] Like Deleuze's metaphysical surface, Freud's psychogeography begins on a frontier. Freud wants to locate, it seems, the "protoplasmic vesicle" not only in space but on the body, at its limit, "on the borderline between outside and inside."

Freud's superficial body lives at the extremities of the outside and inside worlds. It inhabits, but also embodies, those worlds at their limits. "Let us picture a living organism in its most simplified possible form as an undifferentiated vesicle of a substance that is susceptible to stimulation."[46] Over time, Freud speculates, the outer surface has evolved to protect the organism from "an external world charged with the most powerful energies," which, if allowed to penetrate the organism unhindered, would kill it.[47] It has, Freud imagines, gradually become "inorganic," metamorphosed into an immaterial "envelope or membrane" that continues to protect the inside of the organism but no longer has "the structure proper to living matter."[48] This outer layer has become, in Freud's account, an imaginary surface, *Cs*. A trace. The "dead" surface has become a protective filter "against the effects threatened by the external world—effects which tend towards a levelling out of them and hence towards destruction."[49] Consciousness, *Cs.*, is born in Freud's fantastic evolution, in the transubstantiation of an organic membrane. As the vesicle's outer surface dies, it reemerges as a phantasmatic surface, an imaginary tissue that continues to act as a shield against the sheer destructivity of the outside. It is, says Albert Liu, "like a foreskin or callous—a dead protective membrane that can be removed via circumcision or excision."[50] The outside, in Freud's world, represents the forces of a total, one might say nuclear, destruction.

Between exteriority and interiority, corporeal and psychic, material and immaterial, Freud imagines a Borgesean architecture of the system *Pcpt.-Cs.*, an unimaginable place both inside and outside the body, a vital and unique surface, metaphysical, the place of impossible convergences. An archive of surfaces, a surface archive, an architecture of the body that consists entirely of biopsychical layers, in the last analysis, imaginary surfaces. Deleuze names the figure that moves between the surfaces, layers, and planes, the phantasm. It lives at once in a residual and anticipatory relation to the subject, haunting and prefiguring the subject, absolutely singular yet impersonal and divisible—an atomic trace. "What appears in the phantasm," says Deleuze, "is the movement by which the ego opens itself to the surface and liberates the a-cosmic, impersonal, and pre-individual singularities which it had imprisoned."[51] It is marked, says Deleuze, by an "extreme mobility," another feature it shares with the unconscious.

Endowed with a mobility that exceeds the limitations and finitudes of space, the phantasm generates and is itself an unseen energy that swallows space. A "*superficial energy*," says Deleuze, guides the phantasm in the world.[52] Of the mobility particular to the phantasm, Deleuze writes:

> The phantasm covers the distance between psychic systems with ease, going from consciousness to the unconscious and vice versa, from the nocturnal to the diurnal dream, from the inner to the outer and conversely, as if it itself belonged to a surface dominating and articulating both the unconscious and the conscious, or to a line connecting and arranging the inner and the outer over the two sides.[53]

Like the Lumières' train, but also the universal Library, Irma, Berthe, and other figures of impossible corporeality and movement through space—figures for the movement through an imaginary configuration of virtual space—Deleuze's phantasm, drawn from Freud, makes possible such passage. The passage itself is phantasmatic, like Freud's dream voyage into Irma's psychic body, like the "phantom train rides" of early cinema, but it effects the figure of the phantasm as the subject of the passage. That is, the passage into phantasmatic space is made possible by the figure of a phantasm, which it generates in the passage. The phantasm serves as the phantasmatic figure of the phantasm, located on the line—on it but also a figure of it—between interiority and exteriority. A figure of the line.

Railroad tracks and train lines, trajectories forward and backward, vertical and horizontal, the frame line and outline form a series of lines that traverse cinema space, cut and suture the surface (montage), rendering cinema into a series of planes, which expand and contract into and on a metaphysical surface.[54] Lines and figures of lines open and close on the surface. Lines are thus openings *and* closures, sealants and cracks. They are phantasmatic in an unbounded world. The line is a phantasm, the phantasm a line. The phantasm's architectonic structure produces deep surfaces and flat depths, lines and openings, cracks, spaces that turn into and return as other spaces.

The crack, for Deleuze, is neither a sign nor a mark, not even material, but energetic; a movement that establishes a secret opening, temporary and irregular, between inside and outside.

> The real difference is not between inside and outside, for the crack is neither internal nor external, but is rather at its frontier. It is imperceptible, incorporeal, and ideational. What happens inside and outside, it has complex relations of interference and interfacing, of syncopated junctions—a pattern of corresponding beats over two different rhythms. Everything noisy happens at the edge of the crack.[55]

It leaves the surface intact; it is an effect of the surface, phantasmatic, an opening that is not an opening. "What this means," says Deleuze, "is that the entire play of the crack has become incarnated in the depth of the body, at the same time that the labor of the inside and the outside has widened the edges."[56]

"Imperceptible, incorporeal, and ideational," phantasmal. There and not there, the very condition of its *presence* avisual. The crack is formed by surface pressures from within and without. "There is the crack which extends its straight, incorporeal, and silent line at the surface," says Deleuze, "and there are external blows or noisy internal pressures which make it deviate, deepen it, and inscribe or actualize it in the thickness of the body."[57] Phantasmatic and avisual, the crack is an imperceptible opening, a secret entrance into the archive, to the other side, like the dream portal that Freud discovers to Irma's inside. Psychoanalysis can be said to be a science of surfaces and cracks, a science of cracks in surfaces, a science of phantasmatic cracks in metaphysical surfaces.

In "The Ego and the Id" (1923), Freud extends his description of the superficial body and the pressures from within and without that shape it. Freud says, "A person's own body, and above all its surface, is a place from which both external and internal perceptions may spring. It is *seen* like any other object, but to the *touch* it yields two kinds of sensations, one of which may be equivalent to an internal perception."[58] That is, the surface of the body belongs both to the body's exteriority and interiority, is oriented toward the outside and inside. The body is felt (touched) from within and without, and exists at this frontier. Freud's distinction between the senses of sight and touch reveals an economy of reverberations and cracks: the body is seen, but when it is touched, or to the touch, it yields two separate sensations, one exterior, the other interior. Each touch a crack, a small opening to the other side of the body, to the other body.

From the duality of inside and outside that rests on the body's surface, the ego forms. "The ego," says Freud, "is first and foremost a bodily ego; it is not merely a surface entity, but is itself a projection of a surface."[59] Freud expands this idea in a footnote that appears only in the 1927 English edition. He explains: "I.e. the ego is ultimately derived from bodily sensations, chiefly from those springing from the surface of the body. It may thus be regarded as a mental projection of the surface of the body, besides, as we have seen above, representing superficies of the mental apparatus."[60] The ego, in Freud's account, the "I," is fundamentally superficial; it emerges from the body's surface, as a projected surface that embodies the superficiality of the mental apparatus.

Ego and body are surfaces onto which you are projected. You are a projection, a phantasm; the imaginary place where the ego and body converge, where the inside and out adhere, for a fantastic moment, as Freud says of the unconscious, "timeless." You are this screen, on which the disorder of the unconscious is projected. "Unconscious processes," Freud says, "can only be observed by us under the conditions of dreaming and of neurosis."[61] "A projection of a surface." Unconscious processes are avisual. That is, they operate within various modes of visuality, but are invisible unless projected against a screen—dream or neurotic. Their visibility depends on the existence of a screen. The unconscious appears always on the surface, visible only ever as an effect of the surface. And so you are only ever a surface, destined to be always superficial.

For Freud, the geography and history of the unconscious, its figures and forms of movement, and its modes of visuality are derived from a phantasmatic surface, from the trace of a surface long since vanished, along with the prehistoric organism that embodied it. What remains are spectral screens: invisible, avisual, or metaphysical surfaces on which the entire life of the unconscious takes place. Psychoanalysis offers a theory of material and imaginary screens—figurative, abstract, metaphoric, metaphysical screens. In 1946 Bertram Lewin introduced the notion of the "dream screen."[62] An "empty surface," suspended phantasmatically within the dream architecture, facilitates, says Lewin, the perception of otherwise imperceptible psychic phenomena. The dream screen makes invisible psychic phenomena visible, or at least visual. It establishes an internal cinema. Jean-Louis Baudry explains the function of Lewin's psychic surface from the perspective of cinema:

> The screen, which can appear by itself, like a white surface, is not exclusively a representation, a content—in which case it would not be necessary to privilege it among other elements of the dream content; but rather, it would present itself in all dreams as the indispensable support for the projection of images.[63]

The screen is not itself an element of the dream in the same manner that figures and thoughts are, but rather an aspect of the dream apparatus, an "indispensable support" for the architectonics of the dream. The dream is not only an avisual phenomenon—presented visually although not actually visible—not only a temporal event, but also an architectural structure, a topography, a geography, in the material sense of the word, a world. The screen is its portal, its opening, the architectonic crack that threatens to collapse the order of the world, but also allows the passages of phantasms back and forth, inside and out, over and back. A world of screens and a screen world based on the geology of cinema.

Dulac describes cinema, which she calls the seventh art of the "screen," as "depth rendered perceptible"—that is, projected onto the surface.[64] From this perspective, Irma's mouth, and the whole of her interiority—its white patches and secret orifices—can be seen as a kind of dream screen or surface. Freud had projected an image of the dream screen onto Irma's body; he had rendered her phantom body a screen. Freud was dreaming, perhaps of dreams, of phantom interiorities and atomic subjects. The secret of dreams and a secret dream. He may have dreamed not only of a future psychoanalysis but of cinema and the X-ray as well. For the structures that bind the three phenomenologies of the inside always return to the site of a fragile surface; skin resurfaced as screens, screens as metonymies of skin. "The human skin of things," says Antonin Artaud, "the epidermis of reality: this is the primary raw material of cinema."[65] The surface of things and psyche; the "skin of the film," to use Marks's expression. A metaphysical surface or membrane that envelops three organs of interiority—psychoanalysis, X-ray, and cinema. A common epidermis. Deleuze says of "membranes":

> They place internal and external spaces into contact, without regard to distance. The internal and external, depth and height, have biological value only through this topological surface of contact. Thus, even biologically, it is necessary to understand that "the deepest is the skin." The skin has at its disposal a vital and properly superficial potential energy. And just as events do not occupy the surface but rather frequent it, superficial energy is not *localized* at the surface, but is rather bound to its formation and reformation.[66]

"The deepest is the skin," says Deleuze. Its "superficial energy," its "unseen energy swallowing space," is bound to the formation and reformation of surfaces, of the surface, the metaphysical surface. The surface, screen, and membrane form a phantasmatic anatomy, a superficial biology, a biology of the surface. Tracing the genealogy of the dream screen, Lewin likens it to the surface of the mother's body, retained by the dreamer as a mnemic trace. For Freud the psyche and all its figures and phantasms originate on the surface of a now extinct organism. The French *pellicule* already carries this fusion of film and skin, photographic and corporeal surface. The convergence of the screen with skin brings the constellation back to the X-ray.

The surface of the X-ray opens onto an impossible topography, a space that cannot be occupied by the subject or object. Or rather, a space in which the subject and object are dissolved into a phantasmatic hybrid or emulsion. An *atopos* at the surface. The Enlightenment project always was, perhaps, an attempt to move toward this surface, to search for that locus in which the subject would be annihilated by the glare of an intensive radiation, one that moves resolutely toward intensity and interiority. The figure that sutures Irma's body to the irradiation of Hiroshima and Nagasaki, brutal violations of the human surface, may be found in the indelible mark of the X-ray photograph. The X-ray may be the exemplary figure that binds the twentieth century to the legacy of Western science and art, the excess image of an Enlightenment sensibility. Deleuze says, "On these surfaces the entire logic of sense is located."[67]

4. An Atomic Trace

n 1951 the abstract painter Willem de Kooning commented on the radical visuality unleashed by the atomic bomb. The advent of atomic light signaled, for de Kooning, the absolute transformation of visual representation.

> Today, some people think that the light of the atom bomb will change the concept of painting once and for all. The eyes that actually saw the light melted out of sheer ecstasy. For one instant, everybody was the same color. It made angels out of everybody.[1]

An atomic visuality, forged in the spectacular visuality of the atomic or A-bomb, an A-visuality. De Kooning's reflection on the atomic detonation and its effect on visual representation is marked by religious excitement and confusion. The sadistic metaphysics of his account, the cruel suggestion of redemptive ecstasy in the monochromatic annihilation, conveys de Kooning's uneasiness in front of the atomic spectacle. His language charts the limits of figuration before the visual event that may have changed "the concept of painting once and for all." "The eyes that actually saw the light," those who witnessed and understood (or were converted), also lost their vision; in the sacrificial logic of de Kooning's passage, the witnesses exchanged their eyesight for a sublime visuality: the eyes of those witnesses "who saw the light melted out of sheer ecstasy." Ecstatic, outside, blinded. The last form of light, perhaps, that anyone needed to see. The last light of history, according to de Kooning, or the light at the end of history.

"For one instant, everybody was the same color." Which is to say that for an instant, there was no more color in the world. The same transcendent colorlessness illuminated everyone. The catastrophic light of atoms suffuses all people in an overpowering light, which stains each individual body with the purer color of colorlessness. In contrast to the dark light imagined by Tanizaki Jun'ichirô, a secret luminosity that emerges from the depths of the Japanese body, atomic light bathes the body from without, erasing the differences of color and hue that surface each human body. An annihilating, catastrophic light renders the world raceless.

The atomic light, says de Kooning, "made angels out of everybody." Everyone is touched, transformed, but no one survives the force of an atomic metaphysics. De Kooning's anxious rhetoric attempts to account for a spectacle that changes the terms of specularity as such. A spectacle in excess of the capacity of any individual to recognize it as spectacle, or even to see it. De Kooning's angelic, wrathful light of atoms suspends for a moment, but also forever, the economies of visibility and visuality—melting in ecstasy the eyes of those who saw, blending all colors into one, and making everyone angels. A phantom temporality that passes in an instant, in a flash; that leaves behind a historicity scarred and haunted, like Chris Marker's protagonist, by an image, an image of time, torn from its place in history.[2] A timeless image of timelessness. It inscribes an end of visuality, an aporia, a point after which visuality is seared by the forces of an insurmountable avisuality. The atomic blast that melted the eyes of angels brought forth a spectacle of invisibility, a scene that vanishes at the instant of its appearance only to linger forever in the visual world as an irreducible trace of avisuality.

● ● ●

At the time de Kooning sought to fix his understanding of the atomic spectacle in words, to develop an idiom for radical and transformative visuality, another examination of invisibility and avisuality was under way in Japan. A series of minor films based conceptually on H. G. Wells's 1897 novel *The Invisible Man,* but more immediately on the prewar and wartime American films that featured the figure of an "invisible man," emerged in the postwar Japanese cinema.[3] In the context of an imposed *and* internalized prohibition against war references, particularly to the atomic bombings, the *tômei ningen* films in Japan suggest an attempt through popular and fantastic genres to explore the conditions of visuality in the aftermath of World War II.[4]

Adachi Shinsei's 1949 *The Invisible Man Appears (Tômei ningen arawaru)*, a film version of *The Invisible Man* set in postwar Kobe, opens with a scientific competition between two young chemists: each believes that he can discover a method of rendering the human body invisible. The difference between their rival projects lies in the logic that informs each conception of invisibility. One scientist proposes to contract the body's molecular structure to the point of complete density: the opaque body will appear invisible through the paradox of absolute visibility, effecting a kind of human black hole. The other seeks to reorient the body's cellular structure so as to allow light to pass through it like a sieve, making the body appear transparent and thus invisible to human sight. Opacity and transparency frame the dialectic of invisibility, establishing the thresholds of the visible body. Moving in opposite directions, the forces of optical density and dispersal arrive at the limits of visibility, at the thresholds of visuality. According to the terms of this film, invisibility is defined as both the absolute condensation of visible matter and, conversely, its diffusion. Total materialization and total dematerialization institute the same crisis in visuality.[5] At stake in the competition is a woman's hand: the victorious scientist will earn the right to marry the daughter of the two mens' mentor, who supervises the laboratory. (The senior scientist proposes a prize for the winner; the younger scientists immediately suggest their mentor's daughter, Machiko. Just as quickly, he agrees. The inverted oedipal exchange suggests the fluid economies of sexuality and science circulating in the film's diegesis.) The eros that fuels their contest foreshadows the inevitable convergence of light, death, and *jouissance*.[6]

A half century earlier, another woman had offered her hand for the realization of a scientific experiment and became an emblem for transparency. Berthe Röntgen's x-rayed hand in 1895, marked by the exteriority of her wedding band, signaled the entry of light into the human body and the illicit marriage, as it were, of radiation and photographic culture. In the late-nineteenth- and mid-twentieth-century narratives of radiation, the desire to probe the secrets of visual order determined an uncanny rapport between visuality and sexuality, science and art, light and darkness, fantasy and power. The destructive effects of those radiographic histories are embodied in the figure of Dr. Nagai, the doomed protagonist of Ôba Hideo's 1950 film *The Bell of Nagasaki (Nagasaki no kane)*. He is *already* suffering from radiation poisoning, contracted from his overexposure to

X-rays, when the atom bomb destroys Nagasaki on 9 August 1945. Another palimpsest appears in the untraceable X-ray image that opens Kurosawa Akira's 1952 film *To Live (Ikiru)*. As Mitsuhiro Yoshimoto notes, this image, this view of human interiority has no origin in the film's diegesis. No proper history or referent. Yoshimoto calls this floating, unbound X-ray, an "impossible image."[7] Or again in the X-ray film that follows the credit sequence of Teshigahara Hiroshi's 1966 film *The Face of Another (Tanin no kao)*. The protagonist, Okuyama, whose face has been disfigured in a laboratory fire, is seen in an X-ray. Referring to the accident that has left him faceless, Okuyama says: "Everything is too particular. If this had been an effect of human destiny or caused by a war wound or something of that nature, then I might still have been saved."[8] The lethal force of X-rays is recapitulated by the atomic radiation, which echoes the capacity of catastrophic light to penetrate the body and erase the distinction between inside and out, body and environment, images of destruction and unimaginable destruction. X-rays and atomic radiation are linked in a secret narrative, bound by a logic that is historical, overdetermined, and destined—and, at the same time, incidental, accidental, and arbitrary. Wells brought forth his invisible man from the shadow of the X-ray in 1897; he makes explicit reference in his novel to the "Röntgen vibrations."[9]

Adachi's *Tômei ningen arawaru* conveys in its title the paradox of invisibility and transparency as positive modes of visuality. Invisibility functions not as the negation of visibility but as a form of visibility given to be seen, but unseen. Visual but invisible. The transparent or invisible man arrives and *appears* (*arawaru* carries both meanings), suggesting a visuality of the unseen, the arrival of a form of invisibility located within the spectrum of visibility. An invisibility or avisuality that takes place within the frames of the visible, as the condition of possibility of the visual as such. The title implies a semiotics of avisuality, a mechanism for rendering the very invisibility of the invisible at the center of the visible world, to paraphrase Trinh T. Minh-ha.

As the two scientists begin their race to the thresholds of the visible world and the erotic lure that signals from the other side, they learn that their mentor-patriarch has already developed a formula for effacing the human body from the visible world. He has chosen to keep his discovery secret until he finds an antidote to reverse the effect of invisibility. The elder scientist has elected to pursue total transparency rather than opacity and makes clear his preference for the young disciple who has chosen the similar route. The father-scientist's choice of an intellectual heir and future son-in-law tints the dialectic of invisibility with a faint but distinct metaphysics of light. In the context of the film, transparent luminosity comes to be aligned

An Atomic Trace

▼

84

with figures of cleanliness and propriety, while opaque density comes to exemplify those of obsessive ambition. (An argument against the concept of absolute density states that even if the body became invisible under such conditions, it would still cast a shadow. Absolute condensation would result in a shadow without a body, a residue of corporeality that stains the invisible body. A shadow of invisibility, and invisibility as shadow.) Several years after the American film industry experimented with the representation of invisible beings and four years after the atomic bombings of Hiroshima and Nagasaki, which ended World War II, Japanese film audiences were exposed to this attempt to configure a phenomenology of the transparent.

As the film unfolds, its narrative disrupts the contest between the rival modes of invisibility. Members of a criminal organization kidnap the senior chemist and then his disciple, Kurokawa, forcing the latter to ingest a stolen dosage of the compound. He vanishes. With his image held hostage by the underground organization, the transparent scientist turns to a life of crime. A side effect of the potion, disclosed by its inventor, is a proclivity for violence and aggression *(kyôbô)*. As the forces of invisibility begin to consume Kurokawa's existence, he develops symptoms: vengeful rage and jealousy, and the desire for power. Kurokawa returns to the visible spectrum at the moment of his death, after threatening his enemies with a euphemism for his own condition, "I'll erase you from this world forever." Kurokawa is in both senses atomic, an atomic force and dispersed; he is himself an atomic weapon. His rival's research—his search for a material superdensity—might have reversed the effects of transparency by providing a way to shade his transparent body, forcing it back to the spectrum of visible matter. One mode of invisibility counters the other, an antidote invisibility or invisible antibody. Kurokawa dies in the ocean, his bodily form slowly resuming its shape in the water. His death is wrapped in a sublime glow: the overexposed glare of the sun suffuses the image and shimmers on the water's surface. Kurokawa returns to the world underwater, in a ray of luminosity. Kurokawa's once invisible figure has become what Daniel Tiffany calls "the radiant body: the body whose radiant and volatile substance is disclosed only by a nuclear event, the body disappearing in the catastrophic medium of the atom."[10] The formula for invisibility is represented in the 1949 film as a liquid, which is consumed orally. In 1954, when Oda Motoyoshi introduced a new version of the invisible man, *The Invisible Man (Tômei ningen)*, the liquid became a ray and the allusion to World War II moved from an oblique to a direct reference.[11]

The year 1952 marked the end of Japan's occupation by Allied forces and the end of one form of political censorship; a new cinema had begun to emerge by 1954. Between the two films, invisibility assumes two distinct forms: one political, the other phenomenological. Because representations of and references to the war were restricted during and after the war, first by the Japanese government and then by the occupying forces, Japanese artists and intellectuals adopted a variety of rhetorical strategies to address the war and its aftereffects. In the case of postwar Japanese cinema, one finds a consistent recourse to allegory, which determines a representational space otherwise remarkably void of war references.[12] Beyond the political restrictions that shrouded the war, the subject of atomic radiation and its lingering effects in Hiroshima and Nagasaki posed another layer of complex avisuality. The bombings that ended Japan's imperialist activities had introduced a form of invisible warfare or, rather, a form of warfare that circulated through a dense matrix of visuality, displacing any access to a stable referent. At Hiroshima, and then Nagasaki, a blinding flash vaporized entire bodies, leaving behind only *shadow* traces. The initial destruction was followed by waves of invisible radiation, which infiltrated the survivors' bodies imperceptibly. What began as a spectacular attack ended as a form of violent invisibility.

The movement from the American to Japanese films effects several significant changes, including the translation of the word *invisibility* as *transparency (tômei)* in the Japanese versions.[13] The Japanese language has a word for invisibility, *fukashi*. In Murayama Mitsuo's 1957 film *The Invisible Man Meets the Fly (Tômei ningen to hae otoko)*, the distinction between the terms *tômei* and *fukashi* is repeatedly underscored.[14] A scientist developing a powerful ray repeatedly corrects uses of the term *tômei kôsen* (transparent ray), insisting on the word *fukashi*. Despite this and almost by

An atomic trace.

default, *tômei* comes to mean invisible. While both terms imply a diminished form, the nuances of each type of imperceptibility vary. Invisibility suggests a range of phenomenal states, from a material dispersion to radical absence. It implies a metaphysics of the body, an absence at the very core of one's presence. With transparency, the body is there but traversed—violated, like Daniel Paul Schreber's body, by a driving radiance.[15] The rhetorical difference is meaningful when placed against the historical backdrop that separates the American and Japanese films—the war between the nations and the atomic bombings that ended it.

James Whale's 1933 film *The Invisible Man* similarly features two young scientists, a senior scientist and his daughter, Flora, and a love triangle between the brilliant scientist, his lesser counterpart, and the daughter, who is attracted to the radiance of genius. The invisible man, Jack Griffin (played by Claude Rains), has experimented with the dangerous (and fictitious) substance "monocaine," derived from a plant in India, which draws out color and also induces madness.[16] Because the invisible body retains its other physical functions, it is susceptible to the environment, to forces of exteriority such as weather and to the ingestion of food and liquids. Griffin explains that he is vulnerable to rain, mist, and smog; to "smoky cities," "dirty fingernails," basically, ash, dust, and cinder; and especially after meals. A brush against the elements, which adhere to the surface of his skin, and an exposure of the complex dynamic between inside and outside, realized by eating, undoes the effect of invisibility. Invisible corporeality depends, in this instance, on the suspension of normal relations between inside and outside. In this state, the body disappears between worlds, existing neither within nor without the world. Exposed by the encounter with exteriority, which stains the body's surface, and with interiority, which reveals a form of deep invisibility, the invisible body is worldless, otherworldly, between the material and phantasmatic worlds of representation. The living, invisible body remains suspended in this realm of avisuality until death. In a trope that recurs throughout the invisible man phantasm, only death restores the body to a state of visibility. At the moment of death, the body returns to the visible world. Each death scene is almost always preceded by a final representation of invisible movement, tracked by footsteps on the ground, followed by a collapse, death, and the return of the visible human figure.[17] In Whale's film, Griffin returns to the visible spectrum inside out: first his skull returns, his skeletal interiority, followed by his exterior surface. An X-ray image mediates the return from invisibility to visibility—"a death in reverse," says Michel Chion.[18]

Wells's novel adds further details to the conception and staging of an invisible body. In the original, "the stranger," as yet unnamed and unidentified, reveals himself to a frightened crowd. Wells describes the first exposure, the presentation of the invisible, avisual figure before a crowd of hostile spectators. "'You don't understand,' he said, 'who I am or what I am. I'll show you. By Heaven! I'll show you.' Then he put his open palm over his face and withdrew it. The centre of his face became a black cavity."[19]

The confusion between "who I am or what I am" constitutes the crisis initiated by invisibility; in this instance, identity is absorbed by visuality, and invisibility determines ontology. At the center of the crisis is the absent face, the "black cavity," which marks the space of the stranger's being—the obscure empty space of his face. The human face, the exterior surface of the body and metonymy of humanity, serves as the site of a specular avisuality. What is shown but not seen, or seen only as an avisual spectacle, establishes a phantom dialectic that drives the trope of invisibility. "Then he removed his spectacles, and every one in the bar gasped," the narrative continues. "It was worse than anything.... They were prepared for scars, disfigurements, tangible horrors, but *nothing!*"[20] All imaginable horrors are exceeded by the unimaginable horror of invisibility. Effacement surpasses the horror of any material disfiguration. His face fully unveiled, the stranger represents the exemplary spectacle of avisuality: he "was a solid gesticulating figure up to the collar of him, and then—nothingness, no visible thing at all!"[21] He is, says Albert Liu, "a figure of Acéphale."[22]

Among the strange features of Wells's novel is the fact that the invisible man, Griffin, was, in some ways, already invisible, or at least transparent, almost translucent, almost "albino" before his experiments with optical density and chemical invisibility. A virtually colorless whiteness. Becoming invisible seems to have originated in Griffin elsewhere, prior to his decision to seek invisibility. To his interlocutor, Kemp, Griffin begins a scientific explanation of his efforts "to lower the refractive index of a substance, solid or liquid, to that of air."[23] "*Light,*" says Griffin, "fascinated me."[24]

Wells's imaginary science begins with a racist invocation: Griffin relates to Kemp, "I went to work—like a nigger."[25] From albino to nigger, Griffin's transformation across the spectrum of light travels through the registers of race, adding to the idiom of light, the metaphors of racial identity. The *pure* science of Griffin's account is stained by his idiom; his recourse to the language of racism introduces a tremor that runs throughout the novel: the unstable value of the figure of light. Of his initial breakthrough, Griffin states, "I had hardly worked and thought about the matter six months before *light came through one of the meshes suddenly—blindingly!*"[26] As actual light, race, and the metaphors of thought, the trope of light creates a dense rhetoric at the center of Wells's fiction, itself a kind of "black cavity."

"Visibility," Griffin continues, "depends on the action of the visible bodies on light. Either a body absorbs light, or it reflects and refracts it, or does all these things. If it neither reflects nor refracts nor absorbs light, it cannot itself be visible."[27] From this thesis, Griffin introduces the foundation of his discovery, the claim that human beings are essentially transparent. To Kemp's objection, "Nonsense!" Griffin replies:

> Just think of all the things that are transparent and seem not to be so.
> Paper, for instance, is made up of transparent fibres, and it is white and
> opaque for only the same reason that a powder of glass is white and
> opaque. Oil white paper, fill up the interstices between the particles with
> oil so that there is no longer refraction or reflection except at the surfaces,
> and it becomes as transparent as glass. And not only paper, but cotton
> fibre, linen fibre, wool fibre, woody fibre, and *bone,* Kemp, *flesh,* Kemp,
> *hair,* Kemp, *nails* and *nerves,* Kemp, in fact the whole fabric of man except
> the red of his blood and the black pigment of hair, are all made up of
> transparent, colourless tissue. So little suffices to make us visible one to the
> other. For the most part the fibres of a living creature are no more opaque
> than water.[28]

"The whole fabric of man," according to Griffin, is held together, but more important, made visible by the only opaque aspects of the human body: "The red of his blood and the black pigment of hair." In Griffin's account the human body is like paper, a network of fibres and essentially transparent. Another convergence of the body and book (Freud and Tanizaki), the body as book, a book of the body, an archive of the body written, as it were, on the surface of the body itself.

Griffin's attempt to master optical density requires a way to drain the color from blood and pigments. His first success comes when he learns how to change the color of blood from red to white without affecting its functions. A colorless blood. "It came suddenly, splendid and complete into my mind. I was alone; the laboratory was still, with the tall lights burning brightly and silently. In all my great moments I have been alone."[29] Solitude, interiority, and invisibility are bound by the trope of luminosity that runs throughout Griffin's account. "'One could make an animal—a tissue—transparent! One could make it invisible! All except the pigments— I could be invisible!' I said, suddenly realising what it meant to be an albino with such knowledge."[30] Griffin's move toward invisibility, solitude, and madness had already begun; "an albino with such knowledge," he was already withdrawing from the world of visibility.

Even before the completion of his experiment, Griffin's proclivity for secrecy renders him socially invisible. As a "provincial professor," Griffin found it difficult to avoid the constant "prying," which drove him toward greater secrecy and finally disappearance. "And after three years of secrecy and exasperation, I found that to complete it was impossible—impossible."[31]

Griffin needed absolute secrecy; he needed to fuse the social and phenomenal dimensions of invisibility and disappear entirely.

The final task of Griffin's project involves removing the residual color and pigmentation from "the transparent object whose refractive index was to be lowered between two radiating centres of a sort of ethereal vibration."[32] Griffin's compound consists of a mixture of vibrations (radiation) and ingested liquids (strychnine). Between the "ethereal vibrations" of two radiating centers, first a piece of fabric and then a cat are erased, "like a wreath of smoke." But in the case of the cat, not entirely. The experiment failed in two areas. "'These were the claws and the pigment stuff—what is it?—at the back of the eye in a cat. You know?' '*Tapetum.*' 'Yes, the *tapetum.*'"[33] (The *Oxford English Dictionary* describes the tapetum as "an irregular sector of the choroid membrane in the eyes of certain animals [e.g., the cat], which shines owing to the absence of the black pigment.") After some time, the cat vanishes except for its eyes: "After all the rest had faded and vanished, there remained two little ghosts of her eyes."[34] The "black cavity" of Griffin's face has been replaced with the phantom eyes of the animal, which looks at the madman from the vantage point of the invisible. The lines that separate interiority from exteriority, surface from depth, visuality from blindness, and even human from animal being have collapsed. From the invisible world, the cat regards Griffin with only its eyes: with the eyes that remain like the Cheshire cat's grin.

Eventually Griffin subjects himself to the experiment and succeeds. Under the "sickly, drowsy influence of the drugs that decolourise blood," Griffin looks into a mirror and sees his face, "white like a stone." The albino has begun to lose the last traces of his human coloring and begins to suffer "a night of racking anguish, sickness and fainting."[35] Then, after a night, the pain passes.

> I shall never forget that dawn, and the strange horror of seeing that my hands had become clouded glass, and watching them grow clearer and thinner as the day went by, until at last I could see the sickly disorder of my room through them, though I closed my transparent eyelids. My limbs became glassy, the bones and arteries faded, vanished, and the little white nerves went last. I gritted my teeth and stayed there to the end. At last only the dead tips of the fingernails remained, pallid and white, and the brown stain of some acid upon my fingers.[36]

An effect of invisibility, it seems, is a perpetual vigilance, the inability to close one's eyes, since the eyelids, now transparent, no longer block one's vision. Griffin experiences his vanishing body like an X-ray image. The interiority of his body, exposed to his unobstructed gaze, reveals itself before vanishing. An autopsy.

Griffin's inability to arrest his vision follows the general collapse of his surfaces and the lost border between inside and out. As described in the film version, Griffin has to resist eating in public because, he says, "to fill myself with unassimilated matter, would be to become grotesquely visible again."[37] As a result of his invisibility, the inside of Griffin's stomach has become visible to the outside. Anything that enters his stomach can be seen until digested. Griffin's avisuality has turned him, like an X-ray image, inside out. Similarly, his outside, the surface of his invisible body, is vulnerable to the elements. "Rain, too, would make me a watery outline, *a glistening surface of a man*—a bubble," he says.[38] From the exposed interiority of his stomach to the unassimilable exteriority of his skin, Griffin's invisibility has turned him into a contrast of extremes, depth and surface, interiority and exteriority, with no mediation. Fog too traverses his body: "I should be like a fainter bubble in a fog, a surface, a greasy glimmer of humanity."[39] In the withdrawal of the surface, Griffin exists without a balance between inside and out; everything his body comes into contact with remains irreducibly foreign and unassimilable, exposing the invisibility of his body. Contact with the world renders Griffin avisual.

The second and final autopsy returns at the novel's end, on the occasion of Griffin's death. As with most invisible man narratives, the protagonist's body returns to view only after he dies, marking an exchange between life and visibility. In death, the process of invisibility reverses itself: "Everyone saw, faint and transparent as though it were made of glass, so that veins and arteries and bones and nerves could be distinguished, the outline of a hand, a hand limp and prone. It grew clouded and opaque even as they stared."[40] Griffin's body returns to the visible world from his hand, "limp and prone," an emblem, perhaps like Berthe Röntgen's iconic hand, of monstrous visuality and death.

> And so, slowly, beginning with his hands and feet and creeping along his limbs to the vital centres of his body, that strange change continued. It was like the slow spreading of a poison. First came the little white nerves, a hazy grey sketch of a limb, then the glassy bones and intricate arteries, then the flesh and skin, first a faint fogginess, and then growing rapidly dense and opaque. . . . There lay, naked and pitiful on the ground, the bruised and broken body of a young man about thirty. His hair and beard were white,— not grey with age, but white with the whiteness of albinism, and his eyes were like garnets.[41]

The fog flows from Griffin's body, which appears to emit its own weather as it becomes visible. Like a developing X-ray photograph, like Freud's dream of Irma's interiority, Griffin returns to his normal state of transparency, "the whiteness of albinism," an index of his race. Griffin resumes his place in the visible world where he began, with less color than most. White skin and red eyes.

• • •

One effect of invisibility, thematized in the various invisible man scenarios, establishes a relationship between invisibility and madness, and the subsequent desire for power. Power is most often expressed as the ability to cause swift and undetected destruction. In predicting his ascent to global domination, both the film and novel versions of Griffin invoke "a reign of terror." Invisibility, or more precisely the ability to determine one's relation to and place within the visible spectrum, is linked to power, to the possibility of absolute power, which leads to destruction, self-destruction, and ultimately madness. The fantasy of sovereignty triggers the homophone of destruction: from "a reign of terror" to a terror that falls from the sky, like rain, black rain. The sense of total destruction unleashed by atomic war initiated a *fort/da* effect: the closer one moved toward Hiroshima and Nagasaki, the more those topologies receded. At the hypocenter of destruction, a fundamental density left the event invisible. Only its effects, ruined buildings, vaporized bodies, frozen mechanics, and the abstract measurements of lingering radiation (along with other empirical facts—the number of deaths, the heat in degrees at ground zero, etc.) provided an archive of its having taken place, there. Not the destruction of the archive, but an archive of destruction. Like the dialectic between condensation and dispersal, the atomic bombings introduced a visuality of the invisible, a mode of avisuality. They heralded a form of *unimaginable* devastation, in contrast to more recent forms of warfare, which, for distant (televisual) observers, produce *only* images. Instead, the atomic bombings produced symbols— as opposed to images of war—which drove the representation of atomic warfare from fact to figure, toward the threshold of art. The so-called mushroom cloud, which has come to embody the perverse organicity of atomic war, functions as a displaced referent for the obliterating force of atomic weaponry.[42]

Paul Virilio suggests an inextricable relationship between atomic warfare and light, nuclear destruction and photography. "Many epilogues have been written about the nuclear explosions of 6 and 9 August 1945," he says, "but few have pointed out that the bombs dropped on Hiroshima and Nagasaki were *light-weapons* that prefigured the enhanced-radiation neutron bomb, the directed-beam laser weapons, and the charged-particle guns."[43] Weapons of light that introduce new modes of visuality and initiate, like de Kooning's atomic light, crises of visuality. Destructive light and the destruction of light as such.

The phenomenon of invisibility at Hiroshima and Nagasaki has become an essential aspect of its representation in the photographic media. It marks the return of 1895 in 1945, when the discovery of X-rays introduced images of material transparency and fused the function of radiation to photography. What was less apparent in 1895 was the extent to which the new rays facilitated the realization of certain drives intrinsic to photography. The photographic project had always involved more than the mere duplication of nature or the accurate representation of the visible world. Within the depths of what one might call the ideology of photography was a desire to make the invisible visible, but also to engender a view of something that had no empirical precedent. Something never before seen. Like the eruption of four-dimensional matter in a Lovecraft narrative, X-ray photography brought forth, from the depths of the human body, something that had not yet existed—an image of the human body as other, irreducibly foreign, and in its photographic materiality, invisible. Tearing through the opaque materiality of bodies, X-rays transformed photography from an exercise in realism—the production of indexical images—into an allegory of avisuality.

X-ray photography produced a view that exceeded the conventional frames of photography, destroying in the process the limits of the body, the integrity of its interior and exterior dimensions. The body appeared inside out, inside and out, simultaneously. There and not there. Like the invisible men and women of the cinema, X-rays produced a transparent body and reproduced the body as a form of transparency. By passing radiation directly through the body, using it as a kind of radiant filter, X-ray photography exposed the body, in its most intimate, interior depth, as beyond the threshold of visibility. X-rays suggested the arrival of a true scientific revolution, a reorganization of the visible and physical universe.

Not only was the X-ray harmful to the human body, causing extreme forms of sunburn, it actually altered the body's internal structure. This precursor to the metamorphic effects of atomic radiation at Hiroshima and Nagasaki exposed the destructive potential of invisible radiation, but also the photographic properties of the human body. It suggested the essential invisibility contained within the depths of the human body and the photographic image. The symptoms that resulted from overexposure to radiation revealed an uncanny resemblance to photographic processes, suggesting that the body itself could function like a photograph. Thus transformed, the body became a part of the apparatus, absorbed, as it were, by the glare of the photograph.

What was intimated in the radioactive culture of the late nineteenth and early twentieth centuries erupted at full force in Hiroshima and

Atomic shadows.

Nagasaki: if the atomic blasts and blackened skies can be thought of as massive cameras, then the victims of this *dark atomic room* can be seen as photographic effects. Seared organic and nonorganic matter left dark stains, opaque artifacts of once vital bodies, on the pavements and other surfaces of this grotesque theater.[44] The "shadows," as they were called, are actually photograms, images formed by the direct exposure of objects on photographic surfaces. Photographic sculptures. True photographs, more photographic than photographic images. Virilio recounts the photographic legacy of the atomic bombing:

> The first bomb, set to go off at a height of some five hundred metres, produced a nuclear flash which lasted one fifteenth-millionth of a second, and whose brightness penetrated every building down to the cellars. It left its imprint on stone walls, changing their apparent colour through the fusion of certain minerals, although protected surfaces remained curiously unaltered. The same was the case with clothing and bodies, where kimono patterns were tattooed on the victims' flesh. If photography, according to its inventor Nicéphore Niepce, was simply a method of engraving with light, where bodies inscribed their traces by virtue of their own luminosity, nuclear weapons inherited both the darkroom of Niepce and Daguerre and the military searchlight. What appears in the heart of darkrooms is no longer a luminous outline but a shadow, one which sometimes, as in Hiroshima, is carried to the depths of cellars and vaults. The Japanese shadows are inscribed not, as in former times, on the screens of a shadow puppet theatre but on a new screen, the walls of the city.[45]

There can be no authentic photography of atomic war because the bombings were themselves a form of total photography that exceeded the economies of representation, testing the very visibility of the visual. Only a negative photography is possible in the atomic arena, a skiagraphy, a shadow photography. The shadow of photography. By positing the spectator within the frames of an annihilating image, an image of annihilation, but also the annihilation of images, no one survives, nothing remains: "It made angels out of everybody."

Nothing remains, except the radiation. At Hiroshima and Nagasaki, two views of invisibility—absolute visibility and total transparency—unfolded under the brilliant force of the atomic blasts. Instantly penetrated by the massive force of radiation, the *hibakusha* were seared into the environment with the photographic certainty of having been there. In the aftermath of the bombings, the remaining bodies absorbed and *were absorbed by* the invisible radiation. These bodies vanished slowly until there was nothing left but their negatives.

Japan's postwar invisible man films reveal numerous traces of an atomic referent, from overt references to cryptic allusions. In *The Invisible Man Meets the Fly*, the counterpart to the invisible man is a human fly, the *hae otoko*, who strikes from nowhere, perceptible only by the buzzing sound he emits while in flight. The biological experiment that produces "the fly"—which involves a shrinking solution that renders the individual small and light enough to fly undetected in the air—was developed, according to the film, by a Japanese secret weapons unit during the war.[46] The nightclub owner Kuroki, who controls the formula, seeks to exact revenge on those who conducted experiments on him. He has enslaved another man as his "fly," ordering hits on former colleagues in the Imperial Army. During one attack on a woman, the *hae otoko* stalks his victim before descending on her from above and behind. While being pursued by her invisible attacker, she glances anxiously upward looking for the source of the sounds that hover around her. As he strikes her, his shadow and feet appear suddenly, dropped from the sky. "The fly," says Albert Liu, "acts like and is a figure of the ray to the extent that its remote directedness arrives with a sting, a touch."[47] Like the air raids of World War II, the fly assaults his victims from above; the only sign of his presence a distinctly audible hum. The disturbances in visuality introduced by the fly (actually a microscopic, atomic human being) and the invisible man are supplemented by another sense, sound. Avisuality is also heard and should be heard acronymically as an audiovisuality, an AV, avisuality.[48] The metallic faintly industrial sound of the *hae otoko* suggests insects and airplanes, monstrosity and warfare.[49]

In each film invisibility produces tropes of visuality and avisuality, modes of perception in response to the structure of invisibility. Mobile cameras and empty mise-en-scènes, camerawork through vacant space, similarly induce a supplementary sense, even sensuality, in the invisible man films. Another trope is that of blindness: the inability to see with one's eyes, but also as a metaphor of ambition and madness. Of the complex transversality between invisibility and blindness, Wells's invisible man says of his condition, "I felt as a seeing man might do, with padded feet and noiseless clothes, in a city of the blind."[50] An inverse form of invisibility, blindness signals the inability to see what is nonetheless there, somewhere, in the visible spectrum. In Oda's version of *The Invisible Man,* only a young girl, Mariko, who has been blinded in the war, perceives the invisible man, who is himself a victim of medical experimentation by the Imperial Army. Although she cannot see, Mariko can sense his presence in the "air"; she can distinguish the good from the bad, the kind from cruel, each person exudes a unique atmosphere *(me ga mienakutemo, kûki de wakaru no).* Blindness, a frequent motif in postwar Japanese cinema, can be seen as an allusion to the blinding flash, or *pika,* as the atomic explosions were euphemistically called.[51] Eyes melted in sheer ecstasy, as de Kooning says. Like the Greek oracle, the blinded girl cannot see what others see, but can perceive what others cannot. She transcends the physiology of blindness with a form of extrasensory vision. She is, in this sense, visionary.[52]

Of this genre of films, *The Invisible Man* most openly assails militarism in general and Japanese militarism in particular. It seeks to find, like the sacrificial rhetoric of de Kooning, a form of Christian redemption in the atrocity of war. "A truly Christian light," says de Kooning, "painful but forgiving."[53] Nanjô Takemitsu, the invisible man, is a product of the special attack corps and insists that the Japanese military stole his human figure *(sugata).* When he reveals his invisibility to a reporter whom he has befriended, Nanjô says, "Let me show you this model, created by the military state." Unlike Griffin, the bandaged invisible man who unravels, Nanjô effaces himself, wipes his face from the surface of visibility. The human face, says Giorgio Agamben, "is *only* opening, *only* communicability. To walk in the light of the face means *to be* this opening—and to suffer it, and to endure it."[54] Faceless, Nanjô has become an opening without communicability, an abyss. An opening through which only the forces of his affect—anger and faith—pass. Nanjô's anger is tempered by his faith, a theme that runs throughout the film. In the end, he finds peace in self-sacrifice. At the moment of his death, Nanjô returns to the visible world.

Nanjô's face remains a complex trope throughout *The Invisible Man.* He has been effaced (by the Japanese military), has been made faceless (by the radiation), but nonetheless wears a painted face throughout much of the film. He faces the world with and without a face. "The face does not coincide with the visage," says Agamben. "There is a face wherever something reaches the level of exposition and tries to grasp its own being exposed, wherever a being that appears sinks in that appearance and has to find a way out of it."[55] Even a being with no visage can be exposed, can grasp "its own being exposed"; even a being with no visage has to find a way out of its appearance. Nanjô appears and is exposed; he must find a way out of the avisuality that he faces, that faces him, and that constitutes the surface of his invisibility. Nanjô's face is superficial, a pure surface, removable and capable of effacement, invisible and avisual.

Such is the abyss that opens beneath Nanjô's invisibility. On the face and the abyss that it surfaces, Agamben writes:

> Inasmuch as it is nothing but pure communicability, every human face, even the most noble and beautiful, is always suspended on the edge of an abyss. This is precisely why the most delicate and graceful faces sometimes look as if they might suddenly decompose, thus letting the shapeless and bottomless background that threatens them emerge.[56]

Behind every face, but also beneath it, surges the "shapeless and bottomless background that threatens" it. The face is a surface that keeps the formless inside from erupting; it is the surface of formlessness, bound to the formless matter, the formless interiority it covers. Gilles Deleuze and Félix Guattari add: "The face constructs the wall that the signifier needs in order to bounce off of; it constitutes the wall of the signifier, the frame or screen. The face digs the hole that subjectification needs in order to break through; it constitutes *the black hole of subjectivity* as consciousness or passion, the camera, the third eye."[57] The trope of the invisible man creates a disturbance at the very site of signification. Without the screen of the face, the invisible man is rendered pure interiority, transparency, a "black hole." Nanjô's invisibility is a form of interiority, of formless interiority; a mode of formless avisuality. "The only face to remain uninjured," adds Agamben, "is the one capable of taking the abyss of its own communicability upon itself and of exposing it without fear or complacency."[58]

To expose one's face, for Agamben, means to confront the outside, to allow the formlessness of one's interiority to communicate with the surfaces that constitute exteriority. This communication involves a deep commitment to "de-propriation and de-identification": one must become other, you must become other.

> My face is *outside:* a point of indifference with respect to all of my properties, with respect to what is properly one's own and what is common, to what is internal and what is external. . . . The face is the threshold of de-propriation and of de-identification of all manners and of all qualities. . . . And only where I find a face do I encounter an exteriority and does an *outside* happen to me.[59]

The face is a constitutive surface, the limit between inside and out. A *design*. It communicates, or extends a communicability, between the formless invisibility of the abyss—inside and behind—and the exteriority of another. In this sense, the face, which always threatens to "suddenly decompose," represents another surface of avisuality. You are there, this face, facing, but only when you have depropriated and deidentified all that is yours, that is you—when you have moved outside. You are formless when you face the outside, when you are outside.

• • •

Yomota Inuhiko has suggested that the figure of an invisible Japanese soldier may be an allusion to Koreans who were forcibly conscripted into the Japanese Imperial Army and then abandoned after the war.[60] They were subsequently erased from the registers of a visible history. If so, then the Japanese and secretly Korean-Japanese characters are invisible in the sense of Ralph Ellison's nameless "invisible man."[61] Ellison's novel *Invisible Man,* which shares its name with Wells's novel but without the definite article, was begun in the summer of 1945 and published in its entirety in 1952.[62] Some portions of the book were published in 1947 and 1948 in *Magazine of the Year,* fifty years after the publication of Wells's novel. The history of Ellison's *Invisible Man* from 1945 to 1952 coincides with the Allied occupation of Japan.[63] It begins, according to Ellison, with the premise that the "high visibility" of African Americans "actually rendered one *un*-visible."[64] The paradox of avisuality.

From Wells's "albino chemist" who worked "like a nigger" to Ellison's unnamed black protagonist whose invisibility is "a peculiar disposition of the eyes" of others who "refuse to see" him, the two forms of invisibility elaborated by the two novels, set apart by fifty years, extend the dialectic of invisibility to another degree of crisis.[65] "I am a man of substance, of flesh and bone, fibre, and liquids," says Ellison's narrator, "—and I might even be said to possess a mind. I am invisible, understand, simply because people refuse to see me."[66] Alluding to the homonymic figure that precedes him and gives him his pseudonym, Ellison's invisible man explains: "Nor is my invisibility exactly a matter of a biochemical accident to my epidermis."[67] His invisibility is not a material feature of his body, yet it is. His invisibility is entirely an effect of his body, which radiates invisibility, its own invisibility. Like Tanizaki's darkness, which is not a negation of light but a positive and material opacity, the invisible man is constituted visually as invisible; he lives in the visual world as invisible. Ellison's novel, says Fred Moten, establishes "at its heart" a "hypervisibility."[68]

Ellison's avisual figure distinguishes himself from the literary "spooks" and film versions of invisible men, perhaps also from the phantom transparency of film itself: "No, I am not a spook like those who haunted Edgar Allan Poe; nor am I one of your Hollywood-movie ectoplasms."[69] Neither literary nor filmic, the invisible man's world is suffused with light. He lives in a "hole" in New York City: "My hole is warm and full of light. Yes, *full* of light. I doubt that there is a brighter spot in all New York than this hole of mine."[70] A light that fills the hole and enables the invisible man to maintain his invisibility. In contrast to Tanizaki's shadow house, the invisible man's house shines. But the effect is not dissimilar; both houses allow their occupants to see the darkness. The invisible man says:

> I now can see the darkness of lightness. And I love light. Perhaps you'll think it strange that an invisible man should need light, desire light, love light. But maybe it is exactly because I *am* invisible. Light confirms my reality, gives birth to my form.[71]

Light confirms the invisible man's reality and gives him his form: "Without light I am not only invisible, but formless as well."[72] In the depth of his hole, seared by light, the invisible man fights the abyss of formlessness by embracing the very light that renders him invisible. He is a phantasm; his invisibility allows him to evade death and linger, in this world, as a phantom. His is a politics of the phantasm, a politics of deep avisuality that allows the invisible man to follow Clifton and "plunge outside of history."[73] Into a hole through a crack in the surface of the visible world. The invisible man is a projection.

A violent projection that inverts the visual order of invisibility. In the scene of his confrontation with the shadowy white Brother Jack, himself a phantasm perhaps of Wells's invisible protagonist Jack Griffin, the invisible man performs his invisibility, his resistance to the surveillance of others, by disabling the other's capacity to see depth.

> Suddenly something seemed to erupt out of his face. You're seeing things, I thought, hearing it strike sharply against the table and roll as his arm shot out and snatched an object the size of a large marble and dropped it, plop! into his glass, and I could see water shooting up in a ragged, light-breaking pattern to spring in swift droplets across the oiled table top. *The room seemed to flatten.* . . . I stared at the glass, seeing how light shone through, throwing a transparent, precisely fluted shadow against the dark grain of the table, and there on the bottom of the glass lay an eye. A glass eye. A buttermilk white eye distorted by the light rays. An eye staring fixedly at me as from the dark waters of a well.[74]

Brother Jack's face has been disfigured, "disemboweled," the invisible man says; "his left eye had collapsed, a line of raw redness showing where the lid refused to close, and his gaze had lost its command."[75] His eye a projectile, its socket a hole or crack on the surface of his face that leaves his inside exposed. Behind Brother Jack's eye, an opening to the other world that the invisible man has been unable to locate. In this *punctum,* the laws of visuality collapse. Everything is now visible on the surface; all modes of visuality—visibility and invisibility, perception and hallucination, and vision and blindness—are at work in the world rendered monocular, flat: "The room seemed to flatten." A shadow world, flattened, every mode of visuality there, on the surface. And like de Kooning's ecstatic, melted eye, Brother Jack's "buttermilk white eye distorted by the light rays" stares "fixedly at [the invisible man] as from the dark waters of a well." From elsewhere, beneath the surface, the other side of Brother Jack's face. An eye that saw the light of atoms, that has been seared by the force and heat of atoms, by the intense radiation of an atomic figure that radiates a dark invisibility.

If the invisible man is an invisible figure, a figure of invisibility, then he is so as an atomic figure and trace. An infinitely divisible "I" and "I am" and an irreducible trace of the I. His voice an echo of his invisibility, his invisibility an echo of his voice, the phonic materiality of his body. Atomic and anatomic, at once. A visual aurality and aural visuality, like the "flash-boom" *(pikadon)* seen and heard over Hiroshima and Nagasaki. Seen and heard, in some grotesque emulsion, simultaneously and in sequence, one after the other, one inside the other. Aurality inscribed visually, visuality made audible. The invisible man is avisual, audiovisual, in the sense of an impure ensemble, Moten's "ensemble of senses," what he calls the "aural aesthetic."[76] "This aural aesthetic," says Moten, "is not the simple reemergence of the voice of presence, the visible and graphic world."[77] It is rather a crack, a sonic event, which opens onto the visual surface. "How could the silent trace of the incorporeal crack at the surface fail to 'deepen' in the thickness of a noisy body?" asks Deleuze.[78] ("Scream inside and out," says Moten, "out from the outside, of the image."[79]) The invisible man's hypervisibility erupts in the novel—explodes, one could say—as the disembodied voice of the narrator, but also as a sonic boom and blast rendered visual. "How could we not reach the point," Deleuze asks, "at which we can only spell letter by letter and cry out in a sort of schizophrenic depth, but no longer speak at all?"[80] Fractured, divided, schizoid, the invisible man's voice, all of his voices, supplant his invisibility, not as compensation for a deficiency, for a lack of visibility, but as a dimension of his invisibility.[81] A visible voice that can "no longer speak at all." A phonic atomism.

The invisible man describes a scene, shortly after he has been subjected to a series of electrical shocks, in which he is overwhelmed by the "schizophrenic depth":

A man dressed in black appeared, a long-haired fellow, whose piercing eyes looked down upon me out of an intense and friendly face. The others hovered about him, their eyes anxious as he alternately peered at me and consulted my chart. Then he scribbled something on a large card and thrust it before my eyes:

WHAT IS YOUR NAME?

A tremor shook me; it was as though he had suddenly given a name to, had organized the vagueness that drifted through my head, and I was overcome with swift shame. I realized that I no longer knew my own name. I shut my eyes and shook my head with sorrow. Here was the first warm attempt to communicate with me and I was failing. I tried again, plunging into the blackness of my mind. It was no use; I found nothing but pain. I saw the card again and he pointed slowly to each word:

WHAT...IS...YOUR...NAME?

I tried desperately, diving below the blackness until I was limp with fatigue. It was as though a vein had been opened and my energy siphoned away; I could only stare back mutely. But with an irritating burst of activity he gestured for another card and wrote:

WHO...ARE...YOU?

Something inside me turned with a sluggish excitement. This phrasing of the question seemed to set off a series of weak and distant lights where the other had thrown a spark that failed. Who am I? I asked myself. But it was like trying to identify one particular cell that coursed through the torpid veins of my body. Maybe I was just this blackness and bewilderment and pain.[82]

"One particular cell," one atom, that is, particle or building block, inside a body composed of many. The invisible man's identity, his nameless and unnameable identity, unfolds inside a phantasmatic body that is not his. "Maybe I was just this blackness." The blackness that he is, which he plunges into and dives below, the "film without volume which envelops" him, to use Deleuze's phrase, constitutes and bewilders him.[83] He is a phantasm, an avisual phantasm, indistinguishable from the blackness that surrounds him, that he moves into and out of. His voice formed from this blackness, not as the negation of whiteness but as the avisuality of a sonic tremor.

Born in the "summer of 1945," Ellison's unnamed invisible man enters the world at the end of World War II, traversing its final moments and erupting from the shadows of history. From the shadows, a shade that forms on the surfaces of history as an atomic trace. An atomic trace of the collapse of anatomy; a crack on the *anatomic* surface. An end of deep biology. He is a phantasm of a phantasm, a shadow of shadows, who plunges into and out of history, forward and backward into a timeless history, a history without history. At the end of history, the never-ending end of history, of a history without end, infinite and divisible, the invisible man leaves an atomic trace. The invisible man is the figure of this history, its angel. He represents an inverse visuality that arrives with the war's end; a hypervisibility that renders the world blind, for an ecstatic instant *colorless,* making in that photographic moment and *punctum,* "angels out of everybody." An atomic *tRace,* to use Dragan Kujundzic's idiom.[84]

On the eve of the fiftieth anniversary of the end of the World War II, two spectacles scratched the visual surface of the war, the surface of its specularity and spectrality. As the Smithsonian Institution's erasure of the Enola Gay exhibit commemorating the end of World War II rendered the atomic bombing of Hiroshima politically and phenomenally invisible, two avisual echoes of World War II shook Japan in 1995. The Kobe earthquake, on 17 January, and the 20 March sarin gas attack on Tokyo's subway system by members of the Aum Shinrikyō (Supreme Truth) cult projected a shadow of the war at the epicenter of the anniversary.[85] Seen as a return of the repressed atomic bombing, the displaced or deferred spectacles forced the nation to revisit the primal scene of postwar Japan. The Kobe earthquake, Shinoda Masahiro remarks, reintroduced long-dormant images of wartime Japan.[86] The magnitude of destruction sent tremors through the historical and mnemic archives, provoking a nervous anamnesis across Japan. The use of invisible sarin gas by Asahara Shōko, the cult's blind leader, evoked not only the Nazi genocide of World War II and Japanese war crimes but also the threat of an invisible toxin, released into the atmosphere, into the air. The two disasters seemed to force their way into the visible world, returning like memories that were simultaneously familiar and foreign, traumatic, *unheimlich.* Fifty years later, the return of these displaced catastrophic images, along with the Smithsonian Institution's decision to erase Hiroshima, rendered the atomic arena phantasmatic and avisual.

Since 1945 the specter of invisibility has haunted the atomic bombings of Hiroshima and Nagasaki. The unimaginable nature of the destruction has produced a proliferation of concrete and abstract, literal and figurative tropes of invisibility that move toward the atomic referent. The visual materiality of the tropology is marked by erasure and effacement, by a mode of avisuality that destroys the lines between interiority and exteriority, surface and depth, visibility and invisibility. Avisuality is the possibility of the spaceless image, the impossible figure of that which cannot be figured, an image of the very facelessness of the image. It opens onto a site of the atomic spectacle that is irreducibly ecstatic, other — archival. Avisuality is, perhaps, the only true semiotic of the archive. Its only figure, or *sugata*. In the archive of atomic destruction, at its center, in the place where it takes place, inside and out, transparent and invisible, the spectacle of the impossible signifier burns, *cinefied:* radiant, specular, avisual.

5. *Exscription*/Antigraphy

utopsy: "Seeing with one's own eyes" *(OED),* as if in contrast to seeing with another's. (To see one's own death with one's own eyes.)[1] What does it mean to see with one's own eyes rather than another's? How do these two perspectives, these two sets of eyes, alter the scene? Where am I when I see with my own eyes; do I see myself when I look through another's? To see oneself and to make oneself seen—to oneself or another—suggests a remove, a distance, a vantage point on the other side of vision, on the other side of the world, even. One only ever sees oneself or anyone else from the other side. To see oneself one becomes other, an *allopsy.* It is a corporeal project, which involves the projection of a body, of one's body, to the other side. I see myself here, from there. "The enigma is that my body simultaneously sees and is seen," says Maurice Merleau-Ponty. "That which looks at all things can also look at itself and recognize, in what it sees, the 'other side' of its power of looking."[2] Merleau-Ponty's final published essay, "Eye and Mind" (1961), illuminates a crisis: the collapse of the other side,

105

of the screen that divides this side from that, this world from that other one, makes a certain condition of looking, to see and be seen, virtually impossible. With the loss of the other side, I am always here, beside myself. And I cannot see myself here. At stake in Merleau-Ponty's reflection is the possibility of a phenomenology of the self, but even more, of humanity. The humanity of the human body, which he says, is contiguous with the world, which "is made of the same stuff as the body."[3]

Contact between the body and the world makes the humanity of the human body possible. Merleau-Ponty says: "There is a human body when, between seeing and the seen, between touching and the touched, between one eye and the other, between hand and hand, *a blending of some sort takes place*."[4] A blending of the world, of this world and that. The human body, a human body takes shape on the occasion of this contact, this blending of the world. It is an effect of the blending. A mixture of the world and all its stuff.

For Merleau-Ponty, the phenomenology of the human body, the very phenomenon of the human body, is intimately linked to "the problems of painting": "Things have an internal equivalent in me; they arouse in me a carnal formula of their presence. Why shouldn't these [correspondences] in turn give rise to some [external] visible shape in which anyone else would recognize those motifs which support his own inspection of the world?"[5] Painting brings forth a carnal visuality, an embodied and incarnate image, by establishing the internal equivalent ("in me") of the outside world, which is made of the "same stuff." I am an extension of the world, but the world extends, intensifies, forms a "line of intensity," to use Gilles Deleuze's idiom, inside me. The world forms a "strange system of exchanges" with me; I am constituted in an exchange with the world.[6] Painting makes this continuity visible, is itself the visualization of this continuity, of this blending of the inside and out. Images — "designs" and "paintings" — says Merleau-Ponty, are "the inside of the outside and the outside of the inside, which the duplicity of feeling makes possible and without which we would never understand the quasi presence and imminent visibility which make up the whole problem of the imaginary."[7]

"Painting celebrates no other enigma but that of visibility. . . . It gives visible existence to what profane vision believes to be invisible."[8] Merleau-Ponty's painting inhabits the same rhetoric as early cinema: it makes the invisible visible, or rather it makes visibility visible; it forms from the thresholds of the visible and invisible world, an order, mode, or aesthetic of visuality. Not only of the small or fast, but of visibility as such. The visuality of the visible and the invisible is found in the mixture of the body and its world, of your body and your world, all your worlds, all your bodies in this world and all those others. Painting is the process by which the visuality of the visible and invisible is made manifest: "Painting mixes up all our categories in laying out its oneiric universe of carnal essences, of effective likenesses, of mute meanings."[9] Each painting is a universal archive, a picture of the universe, a universal image—and like a dream.

• • •

It is dark disaster that brings the light.
—*Maurice Blanchot*, The Writing of the Disaster

"The Japanese complexion," says Tanizaki Jun'ichirô, "no matter how white, is tinged by a slight cloudiness."[10] Tanizaki says in his treatise on shadows, *In Praise of Shadows*, that Japanese skin actively radiates darkness. Of race and skin color, Tanizaki says:

> From ancient times we have considered white skin more elegant, more beautiful than dark skin, and yet somehow this whiteness of ours differs from that of the white races. Taken individually there are Japanese who are whiter than Westerners and Westerners who are darker than Japanese, but their whiteness and darkness is not the same.[11]

He recalls the experience of seeing mixed groups of Western and Japanese women in Yokohama, noting that despite the careful attempts by the Japanese women to cover their faces and bodies with powder, "they could not efface the darkness that lay below their skin."[12] Beneath the surface, Tanizaki's darkness is profound but apparent, inherent. "It was as plainly visible as dirt at the bottom of a pool of pure water," Tanizaki says. "Between the fingers, around the nostrils, on the nape of the neck, along the spine— about these places especially, dark, almost dirty, shadows gathered."[13] Tanizaki concludes his ironic physiognomy of race by asserting: "Thus it is when one of us goes among a group of Westerners it is like a grimy stain on a sheet of white paper" *(hakushi)*.[14] The Japanese body is an anti-body, a shadow, in Tanizaki's words, an ink stain. Skiagraphic, like the avisuality that Ralph Ellison's "invisible man" projects onto the world. It writes on others, infects them with darkness and invisibility. "My own body's 'invisibility,'" says Merleau-Ponty, "can invest the other bodies I see."[15]

Tanizaki has essentialized the nature of shadows, removing them from the dialectics of light. Shadows, and more specifically the darkness that they signify, are no longer an effect of the absence of light, an interference in the passage of light, but rather an autonomous luminous condition. What Samuel Weber calls "a bright shadow."[16] The transcendent darkness that emanates from Japanese skin is linked, in Tanizaki's discussion, to an essential feature of the body. It seems to originate from within the body, from "beneath the skin," Tanizaki writes, a submerged force that moves toward the surface. The darkness effects an inverted X-ray: a shadow projected from the inside outward, onto the body's surface. Martin Heidegger says, "Everyday opinion sees in the shadow only the lack of light, if not the light's complete denial. In truth, however, the shadow is a manifest though impenetrable, testimony to the concealed emitting of light."[17] Of the classical women who resided ghostlike behind screens, Tanizaki speculates: "Might not the darkness have emerged from her mouth and those black teeth, from the black of her hair?"[18] Irma in reverse, inside out. This genetic darkness is not restricted to the surface of the dark body but radiates outward: it shades not only Japanese skin but extends beyond it, darkening surrounding bodies like black ink on white paper. It pours outward from the body's orifices and extensions, issuing a darkness that functions like a light, a stain, a form of coloring. Tanizaki concludes: "A sensitive white person cannot but be upset by the shadow that even one or two colored persons cast over a social gathering."[19]

For Tanizaki, darkness flows and overflows from Japanese being like a paradoxical light, a liquid light, engendering not only an aesthetic of shadows but also a form of writing. Tanizaki's shadows write not with light but against it—not a *photography* but its antithesis, a *skiagraphy*. They are inscriptions that mark the surfaces of foods, materials, skin, and space itself, assembling throughout the Japanese world an opaque, semiotic empire. Shadows are, in Tanizaki's thought, *expressive*.

Tanizaki's attempts to express a national and nationalist sensibility traverse the registers of light, race, and aesthetics. "Why should this propensity to seek beauty in darkness be so strong only in Orientals?" he asks, adding: "Pitch darkness has always occupied our fantasies."[20] Shadows are not, like the clarity and exposure preferred by Western sensibilities, imposed from outside, but rather originate in an essentially Japanese inclination etched onto the physical features of their bodies. Shadows are, for Tanizaki, originary and inherently corporeal. The Japanese (or Oriental) body is the innate source of the Japanese fascination with shadows. Throughout *In Praise of Shadows*, Tanizaki returns consistently from his wanderings in bathrooms and other dark theaters to the primal site of his discourse on shadows, the Japanese body. "The white races are fair-haired, but our hair is dark; so nature taught us the laws of darkness, which we instinctively used to turn yellow skin white."[21] The body establishes a graphic order, is itself organized graphically, serving as a writing surface. Darkness constitutes and inscribes the Japanese body, which is capable, in turn, of darkening the world around it.

Slightly more than a decade after Tanizaki's reflection on shadows, another form of radiography claimed the Japanese body. The atomic bombing of Hiroshima and Nagasaki in 1945 initiated a new phenomenology of inscription, testing the capacity of the human body to sustain the searing force of atomic radiation. A singularly graphic event, an event constituted graphically, which put into crisis the logic of the graphic. Following Tanizaki's rhetoric of the body as a fantastic surface, atomic irradiation can be seen as having created a type of violent *photography* directly onto the surfaces of the human body. The catastrophic flashes followed by a dense darkness transformed Hiroshima and Nagasaki into photographic laboratories, leaving countless traces of photographic and skiagraphic imprints on the landscape, on organic and nonorganic bodies alike. The world a camera, everything in it photographed. Total visibility for an instant and in an instant everything rendered photographic, ecstatic, to use Willem de Kooning's expression, inside out. The grotesque shadows and stains—graphic effects of the lacerating heat and penetrating light—the only remnants of a virtual annihilation. Virtual because, as Jacques Derrida says, the atomic bombing did not effect a total, irreversible destruction: it did not, to use his phrase, "destroy the archive."[22] Following Derrida's logic, the atomic destruction at Hiroshima and Nagasaki is haunted by the specter of total war, of total destruction. By the shadow of total destruction.[23] Under the shadow of annihilation only the trace remains, a phantasm of the archive, haunted by its own writing. In the remainder, a dark writing was born. A secret writing, written in the dark, with darkness itself. In the atomic night and on the human surface, a dark, corporeal language appeared. At Hiroshima, then Nagasaki, the human figure served as the site of an impression whose syntax defied the conventional modes of understanding. That is, the atomic inscription remained, and still remains, largely illegible.

Against the proliferation of signs they initiated—the various idioms and symbols that have come to stand for the unrepresentable event itself—the atomic blasts also caused a fissure in the corpora of language and signification. On writing in the space of disaster, Maurice Blanchot speculates: "To write is perhaps to bring to the surface something like absent meaning."[24] Disasters do not, for Blanchot, annihilate meaning ("The disaster ruins everything, all the while leaving everything intact"), they render it atomic, inaccessible, secret.[25] The atomic bombings destroyed a certain order of language, a flow of meaning, and forced the topology of language to undergo a radical mutation. Atomic writing, unlike the shadow script described by Tanizaki, does not originate in the body; nor does it arrive, strictly speaking, from the outside either—it comes from nowhere and is, in this sense, *atopical.* Corporeal and atopic, an atopic corporeality: an atomic anatomy. Just as Tanizaki imagines the shadow to originate from within, atomic writing only appears to arrive from the outside. The surface on which both scripts are formed—the human skin—is a tissue that erases the boundaries between inside and outside. Everything that happens on the skin's surfaces represents an unresolved encounter between interior and exterior elements.

Tanizaki's image of an interior darkness spreading outward and staining the Western other like black ink on white paper returns in 1945 as black rain. In his 1965 novel *Black Rain (Kuroi ame),* Ibuse Masuji approaches the unfigurable event through a symptom, one of its aftereffects. Yasuko, Ibuse's ill-fated protagonist, describes the black rain that fell from the irradiated Hiroshima sky. In a diary entry from 9 August 1945, she writes: "I suddenly remembered a shower of black rain.... Thundery black clouds had borne down upon us from the direction of the city, and the rain from them had fallen in streaks the thickness of a fountain pen."[26] The black rain materializes, makes visible, the atomic violence, serving as a writing instrument that transforms the radiation into a script, Yasuko's body into a writing surface. She finds that the dark streaks had stained her skin: "I washed my hands at the ornamental spring, but even rubbing at the marks with soap wouldn't get them off. They were stuck fast on the skin."[27] The body and world fused together, like Merleau-Ponty's phenomenology, in a painting. The image of black rain, liquid atomic ash, falling on Yasuko's face in Imamura Shohei's 1988 film adaptation of Ibuse's novel seeks to render the displaced point of contact between the atomic blast and its victims. The secret archive of the atomic referent. As an aftereffect of the Hiroshima bombing, the liquid inscriptions remain on Yasuko's skin as a visible mark of the radiation—an unabsorbed trace of the violence; an emulsion "stuck fast on the skin." The blending of the world and body, which for Merleau-Ponty makes the human body possible, is not complete on Yasuko's body. Her body cannot absorb this dark toxin; the humanity of her body is suspended on the surface. The survivors of the atomic bombings experience this suspense, says John Treat, "forced to live in a compromise state of both life and death at the same time," dead and alive, in-between.[28] Yasuko's stain is a temporarily visible sign of the radiation that will eventually destroy her. Inscribed on her skin as an initial sign of violent exteriority, the mark of radiation eventually vanishes into Yasuko's body, remaining in it as the imperceptible origin of her sickness. A design. Atomic radiation rendered avisual; first on the surface of Yasuko's body, then inside. There but invisible, radiating darkness. Yasuko has interiorized the black rain; she has brought the world inside her. She has become an environment inside out, radiating darkness, like Tanizaki's Japanese body, outward from within. Black rain pours from Yasuko's body.

Tanizaki's fantasy of an essential, interior darkness and the avisual phenomenon of black rain frame a specific topographical problem. In both instances, there is no fusion, synthesis, or sublation at the place where interiority and exteriority converge—only an uneasy stasis. Tanizaki's Oriental physics suggests the impossibility of ever merging the light, or auras, that surround Eastern and Western peoples into a harmonious whole.[29] The black rain, as a literary trope, underscores the impossibility of understanding the bombings of Hiroshima and Nagasaki: it is a signifier that indicates the inability of language to absorb and stabilize the atopicality of atomic destruction. An *exsign* or *design,* no longer a sign, an exterior sign, a sign on and of the outside, an *exscription.* X.[30] A sign that erases or crosses out—antigraphic. Black rain, like Tanizaki's shadows, can be seen as a figure for the limits of language: a form of writing that is, at the same time, not a part of language, unabsorbed, and unassimilated by the archive. An elemental language, wet and impermanent, is absorbed or evaporates, leaving no inscriptions, only traces. It forms an inscription on the skin of a secret archive. The idea that certain elements can never mix (races and cultures, for example) may have already been part of a Japanese self-consciousness prior to 1945; the atomic assaults on Hiroshima and Nagasaki developed that notion into a philosophical crisis. The atomic bombings created a conceptual emulsion.[31] An idea that opens inside and alongside another, that takes place inside another, within the crypt of another, but never blends with the world that frames it.

• • •

An emulsion is formed in the "mixture of two immiscible liquids (e.g. oil and water) in which one is dispersed throughout the other in small droplets" *(OED).*[32] The notion of an immiscible mixture suggests a paradox, a synthesis that remains, in the end, unsynthesized. A synthesis without synthesis. As a chemical action, the principle of emulsion facilitated the advent of photography in the early nineteenth century. By fixing visible light and other forms of radiation on chemically treated photosensitive plates—in the first instances with a silver compound held in suspension in collodion or gelatin—the photograph holds the image between surface and atmosphere, film and air. The photograph is neither absorbed by the surface nor allowed to dissipate into the air. It generates what Vivian Sobchack calls the "compelling emptiness" of the photograph. The photograph, she says, "is peculiarly flattened.... Figures do not seem to *inhabit* space, to dwell in it, but seem rather to rest lightly on its surface."[33] Suspended between two dimensions and arrested in time, the photograph appears as an effect of the interstice opened by the immiscible mixture. A "vacancy," says Sobchack.[34] It marks the encounter between light and liquid on the surface. An active agent in the inception of photography, the material, chemical, and fantastic properties of an emulsion infuse the photographic and filmic unconscious.

The opening shots of Alain Resnais's 1959 film *Hiroshima mon amour* render the idiom of emulsion in the realms of catastrophe and love, history and memory, image and voice, sexual and cultural difference. An immiscible mixture highlights the impossibility of reconciling the disparate experiences of suffering into a unified whole. "You saw nothing in Hiroshima," he says.[35] The impossibility of documenting catastrophe, of putting words to it; the impossibility of visualizing the unrepresentable. On the bodies and between them, an emulsion constitutes Resnais's film and Marguerite Duras's narrative: it forms a film or tissue between the two principal characters, Hiroshima and Nevers. Their bodies are indistinguishable, covered alternately in ashes and fluids. Cinders and rain. The immiscibility of ashes and rain. Hiroshima and Nevers are unable to merge into one another, unable to perform the phantasmatic syntheses of love. The substances that cover their skin remain unabsorbed, disappearing from the surface by the third dissolve. Their bodies are writing surfaces of an atomic script, corpses, destroyed from within and without.

The film is driven by and suspended in this emulsion — cultural, sexual, existential — which envelops Hiroshima after the bombing. The two unnamed characters have been erased from their homes, Hiroshima and Nevers. He was in the military in China when the bomb destroyed his city; she loved a German soldier and was punished for her transgression by imprisonment and abandonment. Both bear the scars of displacement, of having been elsewhere at the moment of catastrophe. The catastrophe of missed moments. Their brief relationship can be seen as a symptom of their unease, their inability to stay in the place where they belong. The failed union of their romance across sexual and national divides marks the structure of this film's unresolved syntheses: the failure to reconcile fact, truth, or documentary and fiction, catastrophe, and representation; sexual difference, a difference that opens between any two bodies; and destruction and the archive of destruction. The last moments of *Hiroshima mon amour* are inscribed already in the first moments, inscribed on the skin of two bodies, lovers and corpses, two unnameable characters and the faceless dead of Hiroshima and Nagasaki. Writing and effacement, two modes of visuality, become indistinguishable in their obscene mixture, avisual.

The desire to transcribe the atomic experience directly onto human skin represents a symptom of the advent of atomic warfare, a defense of the archive preserved on the body.[36] An atomic trope. The body's surface is where many of the atomic marks were recorded, at times with an excessive visibility, at others with an uncanny invisibility. Writing on the body serves as a ritual repetition of the original violence, the act of anarchiving. A ceremonial act that seeks to make visible a form of writing that is not itself always apparent. The sometimes imperceptible atomic radiation determines a mode of inscription that does not hold fast to the surface, a writing that seems to leave traces before they have ever been marks. Radiation hovers between the surface and the atmosphere, impermanent, appearing only momentarily before either vanishing into the body's depths or evaporating into the environment that surrounds the body. Cinema, a medium bound by the logic of surfaces, provides a metonymy of the human surface: screen and skin. The displaced layers suggest a relationship between cinema space, world, and body, which Siegfried Kracauer has linked with the figure of an imaginary "umbilical cord."[37] Film and body: the terms are conceptually bound by a phantasmatic maternity, an emulsion that keeps them at once apart and together. In the postwar era, Japanese cinema served as a site for another urgent convergence, that of fact and figure, photographic and allegorical representation. Like the X-ray, a sign that points inside and outside and destroys the limits between them, an ex-sign or de-sign, the atomic referent returns to a phantom body, a trace, atomic and anatomic. The fusion of fact and figure, or art in the postwar Japanese cinema, produced an array of film artifacts that retain the logic of emulsions: films that explore the encounters between heterogeneous elements and resist the syntheses of a dialectical writing, representation, knowledge. An impossible fusion. "Or maybe," says Albert Liu, "a *fission*, a figure of both the *crack* and the atomic process."[38]

A number of postwar Japanese films can be seen as attempting to represent nondialectical writing. Two scenes, from two different postwar films, register the disturbed semiology of atomic phenomena. These scenes, from Mizoguchi Kenji's *Ugetsu* (*Ugetsu monogatari*, 1953) and Kobayashi Masaki's *Kwaidan* (*Kaidan*, 1964), exemplify the structures of emulsion in postatomic Japanese cinema, emulsion's permeation of the narrative space. Both scenes can be read as allegories of atomic annihilation—the fear of being enveloped then dissolved by unseen forces, the force of the unseen.[39] Swallowed by an intractable and avisual energy. In Mizoguchi's film, an adaptation of Ueda Akinari's 1758 literary text *Tales of Moonlight and Rain (Ugetsu monogatari)*, the scene involves an encounter between Genjuro and his demon seducer, Wakasa. Disguised as a woman, Wakasa has, in the course of her relationship with Genjuro, gradually depleted him of his vitality, pushing him toward his death. Genjuro—who has forgotten his former life, his abandoned family, and dreams—has become a shade. By contrast, Wakasa has increased her presence in the material world, drawing sustenance from Genjuro's life. In the scene, the two subjects have been displaced from their proper worlds, occupying bodies in varying states of liminality. A local priest recognizes the danger, reading the phantom marks on Genjuro's face.

He proposes to save Genjuro by inscribing his body with Sanskrit prayers. According to the Buddhist belief, the holy text will protect Genjuro from the phantom grasp.

In the encounter between the marked Genjuro and the unsuspecting Wakasa, their last, the Sanskrit talisman interrupts the final exchange between the living and the dead. The scene begins when Genjuro announces his intention to depart from Wakasa's mansion where he has been kept. Attempting to persuade him to stay, Wakasa leans forward to take hold of her captive. At the moment she makes contact, Wakasa recoils in pain, burned by the heat of Genjuro's body. The two are separated at that moment by the surface of Genjuro's skin, his living flesh, which has been stained, rematerialized, by the priest's inscription. It burns her, as if it were still smoldering from the inscription. The written word marked on his body intervenes as a kind of screen. If Wakasa can be understood as a metaphor for atomic annihilation, then Genjuro's textured body has countered the atomic force with its own searing materiality. He is marked by a photographic script and force, is himself atomic. Reclaimed by the living word, Genjuro is turned into an emulsion: part living and part dead, part skin and part text, he has become a suspended animation. The entire scene is figured by the rhetoric of emulsion, of suspense. The registers of spirit and body, feminine and masculine, tactile and optical are suspended within the frame of this encounter. Only by marking his skin, affirming the materiality of its surface, can Genjuro recover his humanity. But Genjuro's is a humanity once lost; he is like the *dehumanized Muselmann* of the Nazi death camps, reclaimed by a language lost and regained, but no longer his, no longer proper to him.[40] A language that returns to him from the outside, from radical exteriority. Genjuro regains his humanity by surviving it; he regains a humanity born in its destruction. Not only do Genjuro and Wakasa form an immiscible mixture, Genjuro, who bears the sign of the liquid ink on his flesh, embodies it.[41] The scene concludes when Wakasa retreats into the shadows, unable to overcome the divide that Genjuro's painted skin imposes, and Genjuro lapses into unconsciousness. He awakens the next morning as if from a dream, lying half-naked and alone beside Wakasa's mansion, which has been reduced to ruin during the night. Wakasa had been, for Genjuro and the audience, a sustained hallucination, a film within a film, a cinematic *mise-en-abîme*. Genjuro's experience with Wakasa is emulsified, never resolved as either dream or reality.

The mise-en-scène that suspends Genjuro and Wakasa determines a unique topography, one that can take place only within cinematic space. The immiscible properties that Genjuro and Wakasa represent are neither blended nor eliminated but rather sustained to produce a third space, a unique realm of phenomenality. That third realm is produced not only as a consequence of the narrative but as an effect of the medium. In the vocabulary of cinema, the encounter between Genjuro and Wakasa can be seen from the vantage point of what Christian Metz has called an "impression of reality."[42] Impressions open, for Metz, a space between fantasy and reality. Seen in this light, one can note the physical as well as psychological nuances that resonate in the term *impression:* that is, impressions can be made on the body as well as on the mind. They are material and conceptual. (Aristotle, who is believed to have identified the phenomenon of "persistence of vision," connects the physical and psychological forms of impression in his treatise "On Dreams." According to Aristotle, dreams are the effects of physical sensations that have been impressed directly on the dreamer's body. The corporeal impression is imagined at night in dreams.)[43] Both forms of impression are enmeshed in the scene from *Ugetsu:* cinematically and thematically, one witnesses the spectacle of impression, the representation of Genjuro's liminality. Genjuro has been reduced to a cinematic impression, to a photogram pressed between the surfaces of imaginary and real space. In Mizoguchi's film, the regimes of illusion and impression are never sublated. They remain until the end, suspended in an existential emulsion. While praising *Ugetsu,* Ueno Ichirô notes this disturbance, "regret[ting] that the fantastic and real realms do not blend harmoniously."[44] The divide separating Wakasa's fingertips from Genjuro's skin and the spectator from the screen represents the atopical tissue that holds the emulsion apart and together.

In Kobayashi's *Kwaidan,* also adapted from a literary work (Lafcadio Hearn's 1904 collection of stories), the emulsion suffuses another scene, another encounter between immiscible elements. The encounter between phantom and flesh is similarly mediated by the liquid ink script. Kobayashi's vignette "Hôichi the Earless" begins with the depiction of a battle from the *Tale of Heike (Heike monogatari),* which chronicles the twelfth-century defeat of the Heike at the hands of the Genji clan. The decisive Battle of Danno-ura in the Straits of Shimonoseki marks the site where the Heike "perished utterly, with their women and children, and their infant emperor."[45] A scene of total destruction.

The story of the blind monk Hôichi unfolds at the water's edge. Famed for his musical skills on the *biwa* lute, Hôichi's recitation of the "Battle of Danno-ura" is said to have reduced even goblins to tears. Hôichi's ability to induce tears in monsters signals his anomalous status at the threshold of the natural world. The trope of phantom tears—the tears of phantoms and the tears that are phantom—suggests a liminal economy that moves from blindness to tears to the liquid ink that eventually conceals Hôichi. Following the rhetoric of tears in Augustine and Nietzsche, Derrida notes the relationship between tears and vision:

Now if tears *come to the eyes,* if they *well up in them,* and if they can also veil sight, perhaps they reveal, in the very course of this experience, in this coursing of water, an essence of the eye, of man's eye understood in the anthropo-theological space of the sacred allegory. Deep down, deep down inside, the eye would be destined not to see but to weep. For at the moment they veil sight, tears would unveil what is proper to the eye.[46]

Eyes are destined to weep, not to see. Hôichi's tears can be seen as a displaced form of blindness; his tears are everywhere—in the ocean and eyes of those who listen, in the ghosts reduced to tears. Pure tears, the essential form, perhaps of the formless. Hôichi's tears are exteriorized, they take place in others as an extension of his blindness, of the sacred allegory that destines eyes to weep rather than see. The tears that Hôichi induces in the demons are blinding, like an atomic flash, apocalyptic. "The revelatory or apocalyptic blindness," says Derrida, "the blindness that reveals the very truth of the eyes, would be the gaze veiled by tears."[47]

Hôichi is positioned from the beginning on the threshold of the senses, between vision and sound, the living and the dead, this world and that other world. Surrounded by waves and tears, Hôichi is approached by the spirits of the dead warriors who invite him to recite the story of the battle. Unable to see his patrons, Hôichi believes he has been summoned to a palace. The waves and tears that wash over Hôichi and his audience elicit in *Kwaidan* what Deleuze calls "liquid perception": a displaced "center of gravity."[48] Over several nights, Hôichi performs the epic tale at the grave sites of the Heike warriors.

As Hôichi's nocturnes begin to exhaust him and the signs of his decay begin to appear on his face, an alarmed priest follows Hôichi to his nightly rendezvous with the Heike. Upon confirming his suspicions, the priest pre-

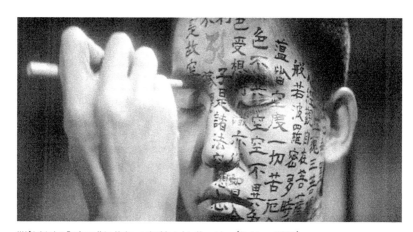

"Hôichi the Earless," in Kobayashi Masaki, *Kwaidan* (*Kaidan,* 1964).

scribes the textual Buddhist remedy: Hôichi's naked body is to be covered with prayers to protect him from the ghosts drawing Hôichi closer to the threshold of death, as Wakasa had done to Genjuro. During the ritual inscription, red paint is first applied to Hôichi's body, merging with the reddish hue that filters the shot, followed by the black ink prayers across the surface of Hôichi's body. The attempt to saturate the body with prayer reveals a material shift from text to liquid, revealing a fantasy of immersion. The fantasy of a liquid immersion in prayer, washed in prayers or displaced tears. (This fantasy also invokes the narrative of a Greek emulsion Achilles, a half-mortal, half-god warrior who was dipped in the river of immortality.)[49] Unlike Genjuro, whose body is marked strategically with the Sanskrit characters, Hôichi is covered in wet calligraphy "on his entire body from head to toe." Immersed in scripture, Hôichi is instructed to remain immobile and mute when the phantom escort comes for him.

The encounter between Hôichi and the phantom warrior begins with the slow materialization of the phantom in the temple courtyard. As its shape solidifies, stopping just short of full density, the phantom starts to call Hôichi's name. The trace of echo, pronounced at first, gradually recedes as the phantom's voice, like its body, begins to stabilize. The phantom's voice has entered the material world of Hôichi. The point of view that determines this scene positions the spectator as its subject. The phantom is rendered from the perspective of a seeing human subject. Receiving no answer from Hôichi, the phantom enters the temple and attempts to locate the blind monk. At this point, the camera captures Hôichi, who is now shown in a semitransparent state. At the same moment, the lighting shifts, revealing not only two systems of visuality but also two orders of light. The shift in perspective transfers the position of the subject to the phantom. Hôichi's invisibility is seen from the unseeing perspective of the phantom. The point of view lingers between two subjects, opening a vantage point that belongs neither to the phantom nor to the spectator. A liminal point of view, an interview.

In the shift from one perspective to another a phantom view emerges, rendering the entire scene transparent. The trope of transparency mediates the dialectic of visibility and invisibility. But the invisible Hôichi is not entirely protected from the phantom's gaze. As the title foreshadows, Hôichi's ears are vulnerable. Having forgotten to cover Hôichi's ears with prayer, the priests have left them accessible to the phantom. Hôichi's ears mark the passage between two worlds of visuality; his body traverses both. His body is visible and invisible, avisual. Seeing only Hôichi's ears, the phantom tears them from his body. The blind Hôichi is rendered earless.

If the blindness, which is often a feature of classical musicians in Japan, can also be seen allegorically as an effect of the atomic flash, the secondary violence to Hôichi's ears at the hands of the phantom soldier reinscribes the memory of war onto Hôichi's body. He is disfigured a second time by the same violence. The trope of atomic disfiguration can be seen in two other examples of postwar Japanese visuality: the case of the so-called Hiroshima Maidens of the 1950s and the figure of an unnamed woman in Teshigahara's 1966 film *The Face of Another,* who hides a scar on the right side of her face beneath her hair. The Hiroshima Maidens were selected from among young Japanese women whose faces had been badly burned in the atomic blasts and were brought to the United States for reconstructive plastic surgery starting in 1955.[50] They became a highly visible emblem of the relationship between atomic radiation and disfigurement. The dialectic of beauty and disfigurement triggered by atomic radiation also appears on the face of Teshigahara's enigmatic woman. In the psychiatric hospital where she works she is surrounded by victims of war and is taunted for the burn on her face, which marks her otherwise beautiful face. (It divides her face into two separate spheres of identity.) On the eve of her suicide, the disfigured woman asks her brother if he remembers the "ocean at Nagasaki," then predicts a war the next day. When the next day comes, she drowns herself in the ocean under a glowing sun while her brother watches in anguish from a window. As his sister walks into the ocean and to her death, the brother screams: at that moment, Teshigahara infuses the shot with what appears to be a ray of light, overexposing the image like an atomic flash. The brother's figure in the window is transformed into a carcass on meat hooks. Facial disfigurement can be seen as one trope of atomic violence, a figure that destroys all figures. The end of figuration in disfiguration. In Kobayashi's film, Hôichi's eyes and ears bear the force of the atomic *pikadon.*

In *Kwaidan,* the optical divide separates flesh from shade, the real from the fantastic, whereas the existential threshold in *Ugetsu* is determined by the tactile senses. The phantom warrior is able to tear off Hôichi's ears but cannot see him, while Wakasa sees Genjuro but cannot touch him. Vision and touch — opticality and tactility — define two modes of perception that mediate a series of oppositional elements: masculine/feminine, life/death, language/body, and natural/supernatural. Opticality and tactility surface in these two films as inadequate means of perceiving the ghostly impression. The two senses converge over the inscribed body, revealing, at that site, their inability to penetrate the phantom world. The nearly fatal encounters in *Ugetsu* and *Kwaidan* chart a space between the two senses, at once optical and tactile, and yet properly neither. Somewhere between the two, a sensual topology begins to take shape. The suspension of the traditional dialectic between opticality and tactility effects an emulsion of the senses, opening the space for more complex sensualities — for impressions rather than perceptions.

The disturbance of the senses in *Ugetsu* and *Kwaidan* is engendered in part by the movement from literary text to cinematic image, by the impression of the written text onto the skin's surface. The relationships of *Ugetsu* and *Kwaidan* to their literary origins are structured to a large degree by their renderings of the transition from literature to cinema, text to specter. Within each scene, the metamorphic force of the literary corpus is figured by the writing that covers Genjuro's and Hôichi's bodies. In both films, the gesture of writing on the body functions as a mechanism for preventing the destructive contact that threatens to absorb Genjuro and Hôichi into the other world of the phantasm. The phantoms can be seen here as figures for atomic annihilation, their mode of being, thanatographic; that is, they force the body to retreat into the secrecy of the archive, but also turn those bodies into archives. The archive inscribed on the surface of the body to render it invisible.

The act of writing on the body in the two scenes suggests a desire to reestablish the absolute separation between the living and the dead, or undead, life and death, this world and that other world. In each case, an *exscription* seeks to clarify the bodies that have become shadows. The *exscription* can also be seen as a symptom of the collapse of meaning, of surface and depth, of the boundaries of existence that are suspended by the atomic radiation. The supplemental tissue intervenes between the liminal corpora, illuminating the profound confusion between the living and the dead in the wake of 1945.[51]

Between literature and film another emulsion forms. Word and image, voice and face are "dismantled," to use Gilles Deleuze and Félix Guattari's idiom; they form a "démontage." The assemblage—Deleuze and Guattari's figure of an irreducible but functioning heterogeneity—resembles an emulsion, by working only through dismantling, through *démontage,* which generates an asynthetic order.[52] Like Germaine Dulac's "decomposition," perhaps, which she posits as the fundamental operation of visuality in film, the assemblage of literature and film in *Ugetsu* and *Kwaidan* produces an emulsion, an "acinema" based on *démontage.*[53] Phantom being overflows into the living world, producing an avisual order that is neither exposed to nor perceived by the senses. It is relayed through secret openings and portals. Both postwar films revisit pre–World War II literary texts that are themselves reflections on war. The literary text itself appears as a trace, a memory, a remnant of the Japanese archive that was virtually destroyed in World War II. The final shot of Ôshima Nagisa's 1976 *In the Realm of the Senses (Ai no Korîda)* captures the crisis: Sada lies alongside her dead lover, Kichi, whom she has dismembered. On his corpse, Sada has inscribed, with Kichi's blood, a declaration of eternal love. A voice-over reveals the film's historical frame, 1936, marking the escalation of the war and its disastrous end. On the eve of an anticipated total war, Sada and Kichi have suspended their impossible consummation, preserving it in the static temporality of a *photogrammatical* instant. A moment marked by liquid inscription. Like Hiroshima and Nevers, Sada and Kichi are frozen in the moment of a catastrophic synthesis: the phantasmatic economy of love cannot suture the logic of a disaster that has, according to Blanchot, already passed and is yet still to come. "To think the disaster," he says, "is to have no longer any future in which to think it."[54] The liquid inscription on Kichi's skin can be seen as a form of protection against the imminent catastrophe, a sign of the graphic events to come, and the impossibility of writing, of archiving disaster.

• • •

If the violent inscriptions of light and shadow on the Japanese body can be seen as a motif for postatomic representation, then the figure of the invisible being may be the logical destination for this trajectory. It is a way to avoid what cannot be seen, or rather, to make that which resists representation invisible. Tanizaki says: "Our thoughts do not travel to what we cannot see."[55] The Japanese invisible man films, postwar adaptations of their prewar American counterparts, introduce a form of *descriptive* subjectivity, a figure of antigraphic visuality. The films can be seen as another attempt to figure the unrepresentable nature of the atomic bombings, an attempt to render visually the radical avisuality of the postatomic world. The idiomatic shift from invisibility *(fukashi)* to transparency *(tômei)* suggests not the absence of visibility but rather the total penetration of the body by light, by invisible rays and fluids. The invisible man *(tômei ningen)* is a figure that remains present, even as it disappears. Or, it is present only in the instance of its disappearance. The familiar trope of the invisible being unraveling—in the case of the *tômei ningen,* erasing—itself inscribes invisibility as a form of erasure. A writing that erases, that produces antigraphic marks.[56] Xs. The transparent being is only there, in the fullness of its invisibility, when it can perform its own effacement. In the idiom of writing, the rhetoric of such invisibility suggests an antiwriting or erasure. The figure's transparency is a form of representational erasure. No longer the effect of a purely interior or exterior force, the *tômei ningen* lingers in the interstices of atomic destruction, occupying a liminal space between life and death, light and shadow, inside and outside. As a trope for the postatomic Japanese body, it is no longer capable of sustaining an inherent racial or ethnic identity. The invisible being is a figure under erasure, but also a figure of erasure, of antigraphy. Blanchot says: "It [the disaster] is what escapes the very possibility of experience—it is the limit of writing. This must be repeated: the disaster de-scribes."[57] The disaster, which cannot be

Stamps from title sequence of Teshigahara Hiroshi, *Woman in the Dunes*
(*Suna no onna*, 1964).

described, itself de-scribes, unwrites, assails the archive by erasing it. Seen
in this light, the figure of erasure initiates a graphic system that annihilates
graphics by destroying the graphicality of the graph. The destruction of
graphicality must also be understood as a form of preservation: it prevents, or postpones indefinitely, the eruption of a catastrophe. The *tômei
ningen*, the figure under erasure, anticipates the possibility of a truly atomic
writing, a writing that can only ever be yet to come, since its arrival would,
as Derrida implies, signal the total destruction of writing itself, of the
future made possible by writing. Atomic writing protects writing as such,
by erasing the text, textuality, and, in the end, the very texture of the body
on which the writing takes place. If an archive of the atomic bombing is to
be built, its history must first be unwritten, *de-scribed* in the rhetorical
economy of an antigraphy: X-ed out.

• • •

Featuring a landscape of fluids and ash, an immiscible mixture of water
and sand, Teshigahara Hiroshi's 1964 *Woman in the Dunes* (*Suna no onna*,
based on Abe Kôbo's 1962 novel) portrays a postatomic world reduced to
ash, reduced to desert. It is set in the outer reaches of postwar Japan, an
inside-out Japan reduced to sand and ash. A shadowless and exposed place, a
fantastic geography burned by a radiant sun. "Of human shadows," says Abe,
"there was not a trace."[58] The film describes a man's gradual disappearance
into the sand, absorbed by its forces, culture, and politics. A politics of sand.
The film, like the novel on which it is based, is marked by the movement

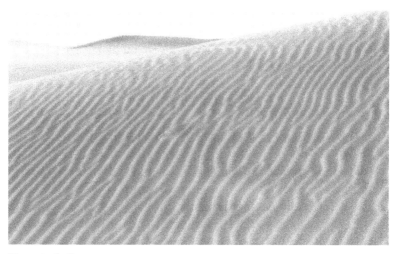

Grains of sand in *Woman in the Dunes.*

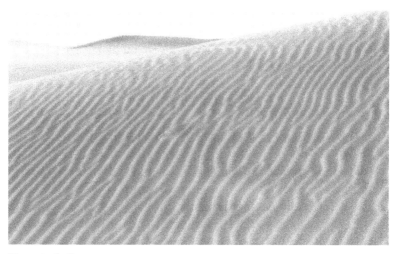

Woman in the Dunes.

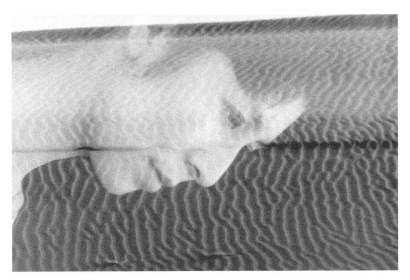

Woman in the Dunes.

from language to image, the written word to an unwritable word; words written in sand. It takes place in the month of Japan's destruction, August, ten years after the end of the war. "One day in August a man disappeared," is the novel's first line.[59]

After the title sequence, which consists visually of official stamps (*hankô*, used as signatures) bearing the names of those listed in the titles, the film opens onto a high contrast close-up shot of a solid, stonelike object. A fade and dissolve turns the shot hotter, brighter, as if burning. The third shot shows a cluster of stones, from a greater distance; they appear translucent, still in high contrast, apparently magnified, giving them the look of irradiated or polarized images. By the fourth shot, from still a greater distance, the stones can be recognized as crystalline clusters of sand, still magnified, still translucent. The fifth shot shows a sea of sand, moving like flowing water; each individual grain has receded and disappeared into the desert. The camera then zooms out, revealing a smooth and animated desert surface, bearing trails or traces of movements in the sand. Teshigahara establishes the homology between water and sand, antithetical elements, in the opening sequence, allowing for the confusion of immiscible elements.

A man enters the scene from the bottom of the frame in a high-angle shot and walks toward the center of the shot, walking away from the camera with his back to it. The camera follows him from behind, his face yet unseen. When the camera moves to the other side, to the front of the man, his face remains hidden by a sunburst, a flash of light that fills the frame and overexposes the shot. The high-contrast cinematography, introduced

8:15 a.m., *Woman in the Dunes.*

in the opening sequence, determines throughout *Woman in the Dunes* a conflict of dark and light: deep, invisible interiority and vast, overexposed exteriority. Secret avisuality and excess visibility.

He begins an interior monologue: "One needs so many certificates: contracts, licenses, ID cards, permits, deeds, certificates, registrations, union cards, testimonials, bills, IOUs, temporary permits, letters of consent, income certificates. Any more? Have I forgotten any?" (The monologue is marked from start to finish by the repetition of "-*sho,*" a suffix meaning paper document.)[60] A close-up of a woman's face, looking downward and in profile, appears superimposed over the desert sand. She turns her head to the right, toward the camera. "Men and women are afraid of a slip-up somewhere."[61] At the end of a slow dissolve between the man in the desert and the translucent woman, who now walks across the desert, she gradually materializes in the shot, in the landscape, like the phantom warriors of *Kwaidan.* The man concludes his discourse on documents: "No one knows what the final one will be. It seems there is no end to them." The woman is now visible in the frame, both of them visible in a two-shot. She is there with him in the desert. The same wind ruffles their hair and clothes. "You say I argue too much," he says to her but in an interior monologue. He is speaking to the woman, who is there, but inside. She is inside and out. "But it's the facts that argue."

His reverie is broken by a voice, "Sensei." Local villagers appear and invite him to spend the night with them, noting that the last buses have left for the day and that there are no hotels nearby. He accepts and climbs

Exscription/Antigraphy

Woman in the Dunes.

down a rope ladder into a deep hole in the sand, to a house at the bottom of a sand dune, where a woman greets him. Inside the dark house, he is fed. One gas lamp lights the interior. A house like the one Tanizaki imagines: submerged in darkness, but forged in sand rather than shadows. He argues with his host, who insists several times that the sand invites moisture and causes decay. He dismisses her views as incorrect and irrational, noting that sand is formed by the absence of moisture; "I've never heard of a moist desert." The dialectic of sand and water, dehydration and liquefaction is suspended in the trope of absorption and immersion: her husband and child, she tells him, were "swallowed" last year by the sand. While he eats, she opens an umbrella over his head to protect him from the sand that falls from above, periodically, like rain. Sand and rain, the absence of fluid and its overabundance, merge in this abyss in the desert. Both are capable of swallowing people. She remains unnamed (referred to sometimes as *kâchan*, slang for an adult woman, meaning "mother," by the villagers); he is called "the helper" *(suketto)* by the villagers, an expression he finds puzzling, and *okyakusan*, or guest, by her. (The actor is Okada Eiji, who plays "Hiroshima" in Resnais's *Hiroshima mon amour* and had earlier, in 1953, starred in *Hiroshima,* Sekigawa Hideo's film about the atomic bombing, excerpts of which appear in the early scenes of Resnais's film.) She implies at several points a longer stay than he plans. He corrects her. His goal is to discover a new species of insect, he tells her, hidden in the remote desert, and to have his name inscribed in an insect encyclopedia. "It's the best I can do." Despite his resistance to identification papers, he wishes to have his name inscribed in an encyclopedia, an archive of sorts. Their first night ends with her shoveling sand during the middle of the night. She does this all night, every night, to protect the nearby village from being swallowed by the eroding sand. Her dune is one line of defense against the consuming force of the sand. He offers to help, but she refuses, saying "not on your first day." He corrects her again. She gathers the sand, then sends it above by pulley.

The next morning begins with a close-up of a watch, seen from his perspective; it shows 8:15 a.m. Teshigahara's evocation of atomic time, the frozen moment of Hiroshima's destruction—absent in Abe's novel—establishes the referent, if not allegory, of Hiroshima in *Woman in the Dunes*. Okada's face and body, inscribed in three films that reference Hiroshima—two directly, one obliquely—become the figure for a transfilmic inscription of Hiroshima and atomic violence. A faint and sustained orchestral sound, mixed with the sound of wind, triggers a series of cuts and dissolves, a montage of images one to another in a sustained rhythm. The face of the watch, secondhand ticking, cuts to a close-up of his face, his illuminated eyes set apart from his otherwise dark face. He turns his head slightly to the right and moves his eyes to the side, then up, and for a moment appears to meet the camera directly before turning away. His absorption becomes momentarily theatrical before it recedes again into absorption. A profile of a woman's body, his host's sleeping body, appears in superimposition over his face. From the right side of the frame, a wipe: the desert moves across her body, the waves in the sand match the curvature of her hips. Sand over the woman's body, the woman's body over sand. The shot echoes the phantom addressee that appears earlier, but also the opening shots of *Hiroshima mon amour*. Bodies written in the sand, written with sand, inscribed in and by it. Sexuality is rendered in ash, as the ashes of desire and passion.

The music reaches a feverish pitch, then suddenly stops. The shot cuts to a medium shot of her midsection, her stomach moving as she breathes. His face returns in profile. He appears to be gazing intently at her. The watch now shows 11:20 a.m., suggesting that he has been mesmerized by the view of her naked, sleeping body for three hours. It is not clear why he has waited so long, hoping perhaps to thank her or, more likely, unable to take his eyes from her body. He has become fascinated by her. As he prepares to leave, he continues to survey her sleeping body, which is covered in sand. She appears to be made of sand, an extension of it; her elemental body formed from the sand that surrounds and constitutes her. The title of the novel and film, *Suna no onna*, is translated as the "woman in the dunes" but means more precisely, "the woman of sand."

He hoists his rucksack onto his back and wipes the sand from his hair. The sand that has already begun to enter his body. He leaves money for her, tucking it under the kettle. A brief shot of the house, seen from above, shows it overexposed and overheated. An aerial view, as if seen from an airplane. Outside, he surveys the barren landscape and notices that the ladder, which he used to descend into the dune, has disappeared. He tries to climb the side of the dune, but the sand crumbles with each grasp. And then he realizes that the ladder can only be lowered and removed from above. He suddenly understands his predicament, her complicity in his capture. He is a prisoner of the villagers and this woman, but also of the elements, of the sand that surrounds him.

His initial reaction is anger and defiance. "I'm a registered citizen." His attempts to escape, to upset the fragile sand structures, only cause the sand to descend on him more violently, in the end, injuring him. Back inside, his next move is to take her hostage and refuse to work. The sand dunes continue to crumble, the work remains unfinished. Teshigahara shows a closeup of her face and neck, the camera traveling along her chin and the base of her neck. The sand stuck to her drying skin appears to seep out of it. As if she is exuding sand from the inside. As if she is turning to sand as she dehydrates. Her interior becoming exterior. Several moments later, Teshigahara repeats the slow movement of the camera across the line of the man's chin. His skin looks like hers; the two are becoming indistinguishable on the surface. He is going mad, losing his identity in her and in the sand. The sand *deforms* the landscape and them; it renders the world formless, reducing everything to the atomic structures of sand.

The earth shakes, and the two collapse onto each other and embrace in a moment of erotic realization. The view of the sand that covers their bodies and suspends them like the statues of Pompeii recalls the opening shots of Resnais's film. It is a quotation of it; indeed, the sand falls on the same body. The storm abates, the raining sand softens to a drizzle. Slowly they rise, self-consciously. She adjusts her blouse and hurries into the next room. He brushes the sand from his pants, removes his shirt, shakes the sand from his hair. He follows her to the next room, where she is changing. Through a translucent screen he asks, "Shall I brush you off?" As he wipes the sand from her body, she slips into a sustained *jouissance,* clutching passionately at his body. They fall to the ground and make love in the sand, invoking again the ash that falls on Resnais's bodies. Outside, sand drips slowly along the side of the dune like a thick liquid.

But the exchange of fluids from body to body does not replenish either body; it accentuates the line between desire and thirst. The drought continues, their madness intensifies. After he hallucinates an oasis, he relents.

Woman in the Dunes.

"My blood will rot," he says. "I surrender, I surrender." He steps into the blinding sun. He lights a rag soaked in alcohol and signals his surrender to the villagers. A superimposition of water soaking the sand. And the water returns, saturates the frame. A large droplet drips slowly in closeup. He gives in to the inevitable forces of the elements, water and sand.

Accepting his fate as a prisoner, the guest restrains his anger and adopts a strategy of patience. "It's just a matter of patience," he says, "until I'm rescued." He begins a series of quasi-scientific experiments and low-grade engineering projects, long-term plans for escape. He continues his projects, collecting insects and building crow traps. Time is passing, slipping away into the sand. When the villagers deliver his rations, he asks for a favor: he asks to be let out once a day to see the ocean, just thirty minutes, twenty minutes is enough, even ten. "I promise I won't try to escape." The villagers promise to consider his request. His rations are now more elaborate and include eyedrops—synthetic tears, perhaps, which maintain his eyesight. The sand is infiltrating his body, entering through his eyes.[62]

The villagers return with an answer to his request. They will grant him his daily walk along the ocean if he is willing to have sex with her before them. A spotlight shines on him from above. *Taiko* drums fill the scene. The entire village, it seems, is assembled above, along the rim of the dune, many of the villagers wearing masks. They look down on him from above in an extreme high angle: a ceremony of self-debasement or effacement that will allow or force him to become one of them. They are masked, defaced. When he appeals to her, she dismisses the insult at once. Lured by the possibility of escape, he assaults her, but is unable to complete the act.

The villagers disperse, the scene exhausted. He has become their spectacle. An insect.

He returns to his traps and discovers one filled with water. He tastes the water, which is fresh. The contraption works like a pump, drawing moisture from the sand. The discovery provides him with a means of self-sufficiency and resistance. He is overcome with joy and immediately buries himself in work, hoping to improve its efficiency. In his dehydrated hole, another hole that draws moisture from the earth. A *mise-en-abîme*, a hole within a hole, in the earth, which reaches to the other side, to life. In the winter, she collapses in pain. She is pregnant, the pain a sign, perhaps, of an irregularity. The villagers lift her from the dune, and he watches her vanish above, into the outside. While the villagers are in his house, he resists the urge to confide his discovery to the village leader. When they leave, they forget to retrieve the rope ladder, leaving him free to escape.

He climbs out of his prison. He walks along the windswept dunes and reaches the ocean. Outside, it is still light. He can escape. In a close-up, he turns his head, panning across the landscape like a camera, surveying it like a geographer. He turns away from the camera and faces the ocean. Cut to a profile, in extreme close-up. He looks out onto the ocean. A shot of footprints in the sand. Then, he is climbing back into his dune, into his prison. He checks his water pump, measures its level. As he peers into it, he notices a figure in the reflection, above and behind him. He turns to look. A boy, peering down at him. He looks upward, the boy withdraws. A glimpse of his future life, perhaps. In interior monologue, he begins to speak: "There's no need to rush anywhere just yet. . . . I'm free to do as I like now. At the moment, his focus is on his experiments, on telling someone of his discovery. After that, "I can think of escaping."

Inside and out, imprisoned and free, lost and found, this man is an emulsion. He operates according to the logic and visual economy of the film: superimposition. One over the other, one inside and alongside another. Two scenes that coexist uncomfortably, never mixing: sand and water, smooth and striated spaces, letters and sand, man and woman, the same man and another woman, this moment and that, this world and another. He is inside and outside her, she of him. This world in that, that world in this. The sand swallows the sand and everything else, each surface collapses onto another.

In the film's final shot, a handwritten document appears superimposed over the dunes. The words appear to have been written on the sand in rigid, striated text contrasting the sand's smooth and unstable surface.[63] The document, a missing person's report, records the name of the protagonist, "Niki Junpei, born 7 March 1927." (Abe specifies the date of the report as 18 August 1955; the events of the narrative traverse the tenth anniversary of

Exscription/Antigraphy

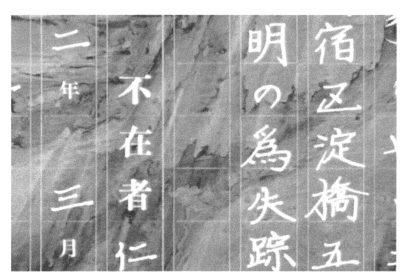

Missing person's report, last shot in *Woman in the Dunes*.

Japan's surrender on 15 August 1945.) Yet another document, the last one. The first time his name is mentioned. It comes to him, his name, at the moment it is no longer his: he is no longer Niki Junpei, Niki Junpei is no longer him. A posthumous name attached to him like text, to his body, which has vanished. His former name now only a series of arbitrary lines, etched in the sand. Of the line that vanishes in the sand, in the smooth archive, losing its figurative and spatial forces, Deleuze and Guattari say:

> *A line that delimits nothing, that describes no contour,* that no longer goes from one point to another but instead passes between points, that is always declining from the horizontal and vertical and deviating from the diagonal, that is constantly changing direction, a mutant line of this kind that is without outside or inside, form or background, beginning or end and that is alive as a continuous variation—such a line is truly an abstract line, and describes a smooth space.[64]

His name has left him, leaving behind a life "without outside or inside." The report concludes that after seven years, he is now considered lost. He has disappeared. An invisible man, neither living nor dead, suspended in an interstice like Schrödinger's cat, an emulsion. The inscription of his name, which he initially resisted and which led him to the desert, he also sought in the form of recognition in the insect encyclopedia. The final documents that erase Niki Junpei from the world invoke the *hankô* seals that appear on the opening credits. Framed by two forms of inscription, Niki Junpei has been inscribed antigraphically in an archive of sorts, an archive of the vanished. He has been inscribed, name and body, in an archive of sand. It is an antigraphic archive, written and erased in the sand. Like thousands of dehydrated atoms, Niki Junpei, identified only on the occasion of his disappearance, has become invisible to the world, lost to language. His name a series of abstract lines in the smooth sand, his body a guest in the desert. He is dispersed in a universe of sand, indivisible from it and invisible in it—avisual and atomic.

6. Phantom Cures:
Obscurity and Emptiness

wo films at the end of the twentieth century reflect a continued disturbance in the visual economy of Japanese cinema, sustained since the
end of World War II. The films, neither one representative of late-
twentieth-century Japanese cinema, are nonetheless exemplary in their
examinations of the fragile relation between avisuality and unimaginable
destruction. Kore-eda Hirokazu's 1995 *Maborosi (Maboroshi no hikari)* and
Kurosawa Kiyoshi's 1997 *Cure* explore thematically and materially the phenomena of superficial obscurity—an opacity that renders the surface
inward—and empty depth, an abyss without volume. Fifty years after the
end of World War II, one hundred years after the discovery of X-rays,
psychoanalysis, and cinema, Freud's unresolved problem of psychic representation, of a psychic visuality (of a visuality specific to the psyche) returns
in two films haunted by a displaced interiority, by the evaporation of interiority into the atmosphere. *Maborosi* and *Cure* offer shadow psyches,
available only as obscure traces of an inside projected against the surface;

an impossible psychoanalysis that is itself the very origin and essence of psychoanalysis. Each work generates a spectacle of invisibility, an avisuality that frames death as a form of opaque vitality.

"Maboroshi no hikari" (the phantom light or light of illusion) refers to an explanation offered to Yumiko, who seeks to understand her husband's unexplained suicide. She is haunted throughout the film by the uncanny repetition of loss: first her grandmother, whose suicide she failed to prevent as a young girl, then her husband, Ikuo, who kills himself one night, suddenly and without explanation, by walking along the train tracks into an oncoming train. In both instances, Yumiko feels suspended between responsibility and guilt, bound to and abandoned by each figure. The phantom light is invoked by Yumiko's second husband, Tamio, as folk wisdom among local fishermen who claim that a light sometimes appears to lone figures at sea and lures them to their deaths in the ocean. Tamio never uses the expression "maboroshi no hikari"; he says "a beautiful light" *(kirei na hikari)*. The film's title and the phantom light are in *Maborosi* nondiegetic, projected inward from elsewhere, from a space outside. Kore-eda's film is suffused with faint phantom lights that enter the dark interiors of *Maborosi*'s world. Natural lighting throughout much of the film establishes a dim luminosity that makes, at times, *Maborosi* virtually invisible: small windows, street lamps, and light bulbs often act as the sole sources of light in dark scenes, shot most often in medium and long shots, maintaining throughout the film a sense of irreducible distance and obscurity. The phantom lighting of *Maborosi* suspends the film in a liminal state, on the thresholds of perception, always at risk of disappearing entirely.[1] Minimal light and visibility tend to condense deep space, rendering the film—its world and characters—opaque and obscure: faint traces on the film's exterior surfaces.

The film opens with a dream, unmarked as such until Yumiko awakens and the diegesis of the film is established retrospectively. Yumiko's dream coincides with the film's first scene. The film fades into a frontal medium shot of a young girl sitting in a dark room looking at herself in a handheld mirror. She hears a bicycle bell and turns away from the camera to look, as a blurry figure passes swiftly behind her, outside. She follows the sound, stands up, and exits the room. Outside in another dark space, she sees a boy riding a bicycle through a tunnel toward the other side. Someone, presumably his mother, calls after him, "Ikuo, come and help." He ignores her and continues. After a brief shot of an empty doorway seen from inside a dark alley, the view returns to the tunnel. A figure walks through the dark tunnel toward a sunlit street at its end, moving away from the camera. An old woman. She reaches the other end and exits onto the street to the left. As soon as she disappears, the young girl enters the tunnel running in pursuit. In the next shot, an elderly woman on an arched bridge turns to face the young girl, who pleads for her to return. "I want to die at home," the old woman says, identifying herself as the girl's grandmother. The girl tries to dissuade her, but the grandmother insists and leaves the girl, walking away from the camera toward the other side of the bridge and her death. The girl stands in the distance, in a long shot with her back to the camera as her grandmother leaves.

The dream, still unmarked, continues. Under a streetlight, a man and two boys light sparklers on the street. The girl, again with her back to the camera, stares outside a window into the dark night. Her father returns, having failed to locate the missing grandmother. "What if she never returns," says the girl. "It's not your fault," her mother responds, identifying the girl as Yumiko. The scene returns to the bridge at night, lit only with street lamps. Yumiko runs across the bridge in the dark in the direction of her grandmother. She stops, her figure barely visible in the dark, at the same point where she left her grandmother. Yumiko walks dejectedly through the tunnel that leads back to her home. She hears a bicycle bell and then sees a boy pushing his bicycle. The two stare at each other, and, as the scene fades to black, a woman's voice asks, with a slight trace of echo, "Iku-chan?"

This voice, marked by the exteriority of an echo and the echo of exteriority, acousmatic and for the moment nondiegetic, serves as a transition from the dream. A tunnel. In the dark, the voice repeats with less reverberation, "Iku-chan?" A man groans. "I had that dream again. I have it often recently. I wonder why," says Yumiko, to which Ikuo responds, "I'm not the reincarnation of your grandmother." Yumiko turns on a light, and the two are visible for the first time in the present. He turns off the light, plunging the scene back into the dark. "Go back to sleep. Keep dreaming. Maybe your grandmother will return." "Maybe you're right," says Yumiko. "But why couldn't I have stopped her then?" In the darkness, from it perhaps, nondiegetic music enters the film, and the title appears, followed by the first shot of the film after the title, a man riding a bicycle toward the camera, on what appears to be the same bridge crossed by Yumiko's grandmother. A reincarnation perhaps, Ikuo returns from the other side.

Yumiko and Ikuo's exchange in the dark invokes the opening lines of Alain Resnais's *Hiroshima mon amour*. Acousmatic voices that pierce a world not yet acknowledged, not yet given. An acousmatic exchange between two phantoms, timeless. The opening of *Maborosi* establishes an atemporal flashback, a flash backward in time folded into the space of a dream.[2] A past outside time, like Hiroshima, a phantasmatic history outside history. A past that falls, like Ralph Ellison's "invisible man," "outside history." A catastrophe, timeless in its endurance, and atemporal in its resistance to the laws of time, to chronology. Yumiko's preliminary voice arrives before her, in advance, before the film's opening: a voice that comes from outside and establishes the impossibility of seeing, of understanding. The impossibility of returning to that voice a visuality and temporality proper to it. "You saw nothing in Hiroshima." The voices have not yet been embodied, they have not yet assumed the corporeality and visibility of individual human beings. They are not yet proper to anyone. They signal the avisuality of disaster, the impossibility of its representation. Past history folded into a dream, displaced in the film from the diegesis proper to it. Yumiko's dream is an element of the film's story, but also a glimpse into an interiority constituted like the phantom lights that pierce *Maborosi,* from the outside. The film is constituted architecturally as an outside interiority, and an inside exteriority: as the impossibility of establishing stable orders of inside and out.

Central to the narrative of *Maborosi* is the illegible suicide of Yumiko's first husband, Ikuo. His disappearance initiates the narrative and haunts the film as its empty center. One night, Ikuo fails to return from work and is discovered dead later that night. In the middle of the night, Yumiko is asked to identify his corpse. Ikuo walked directly, she is told, into an oncoming train, ignoring its warning lights and horns. He provided no explanation, no sign of his intention. "It's as if he left behind a riddle," says Yumiko's mother.

Ikuo's death invokes the fantasy and spectacle of train collisions in early cinema. In a sense, a cinematic death. Images of trains crisscross *Maborosi,* carving lines in the film space that move across the surface of the film and also into its phantasmatic depths. Trains are part of the landscape of Japan, but they also echo the trains and their trajectories that structure early cinema. One hundred years after the Lumières' first films, Kore-eda's trains still establish illusory depths, lines that vanish into phantasmatic distances and those that secure the materiality of metaphysical surfaces, to use Gilles Deleuze's expression. "Unseen energies swallowing space." In a later scene, Yumiko's son, Yûichi, and Tamio's daughter, Tomoko, ride a train through a tunnel, alluding to the "phantom ride" films popular in early cinema. The shot is filmed from inside the train directly ahead. It begins with an abrupt cut from the previous scene. Neither child is yet visible in the train. As the train approaches then enters the tunnel, the frame gradually darkens, leaving only the light at the other end. The scene cuts to a dark frame, the sound of children singing. Barely visible at first in the darkness of the tunnel, the two children gradually appear in the light, shot from behind, looking out from a window on the side of the train. The view into the tunnel has not been established by any of the characters, by any subjectivity located within the film, but rather as a disembodied point of view, a phantom ride.

The geography of *Maborosi* is mapped as a sequence of passages through tunnels, interior corridors, alleys, and mountain roads, and across bridges:

movements from one side to another, to the other side. Yumiko's grandmother passes across the bridge to the other side. A dark tunnel leads to and from the neighborhood where Yumiko and Ikuo spent their childhoods, smaller ones lead from Yumiko and Ikuo's residence when they are adults to their surroundings. A winding mountain road takes Yumiko to her second husband, Tamio, in Notô. Inside, dark corridors connect one dark space to another. Throughout *Maborosi*, characters pass through tunnels on foot, on bicycles, and in vehicles. In Notô, Yûichi and Tomoko run through a dark tunnel toward the bright green light at the other end. Bicycles, trains, automobiles plunge into and emerge from dark tunnels, vanish into distant space. Against the flat surfaces that constitute the interior volume of *Maborosi*, those trajectories that transport the characters into the film's phantasmatic interior establish a spatial order that is exemplarily cinematic. Flat depth: a world rendered superficial, a metaphysical surface. Every passage to the other side, every return from the other side, leads to another surface, another plane. In this way, sequentiality without depth is developed.

The flat volume of *Maborosi* is constructed by a series of planes, a *mise-en-abîme* of flat surfaces extending within flat surfaces, receding into a perpetually superficial depth. Flatness inside flatness, behind and on either side. On the other side, the continuation of the surface. Each passage framed by a threshold: windows, door frames, tunnel entrances produce dark abysses. Rooms behind rooms, seen from a distance, the frames and borders usually an element of the image. Moving usually from one dark space to another. (Kore-eda claims to have used natural lighting exclusively, except in the scene in which the newly constituted family eats watermelon on the veranda.) Planes and surfaces receding in space; behind every flattened space, it seems, another dark surface. An extension of surfaces, one beyond the other, but also an intensity of surfaces that move inward. The superficial structure of *Maborosi* makes a spatial logic of insides and outsides ultimately untenable.

Kore-eda's organization of space in *Maborosi*, his architectonic order, is largely vertical and horizontal, with few oblique angles and diagonal compositions. Almost every shot and everything in it, every object and figure, is framed by a flat opacity. The low camera angles and frontal composition, still figures and long takes suggest a darkly photographic world, strangely immobile, paralyzed, and timeless. Embalmed, like André Bazin's photography. What moves through the film, what infuses vitality into its static life is a phantom drive: the death drive that Ikuo projects into an otherwise quiet world.

Yumiko's return to Osaka where she once lived to visit her mother reintroduces the theme of haunting and phantasmatic life. Each space — streets, coffee shop, workplace, apartment — is marked by an absence, the absence of Ikuo, but an absence already at work when he was alive. The ghosts that linger in Yumiko's past world are reanimated upon her return by the acousmatic sound of the bicycle bell. As Yumiko walks through a familiar passage toward the camera, the sound of a bicycle bell pierces the scene. Yumiko turns in response; a woman bicyclist crosses the plane horizontally behind Yumiko. Someone else. The phantom sound intact, carried by another body. Yumiko returns to the factory where Ikuo worked, repeating an earlier sequence of shots in which Yumiko observes him there. She retraces her steps to Ikuo's workplace. Kore-eda composes a view of Yumiko from inside the factory, a haunted point of view, like the ghost shots of Kobayashi Masaki's translucent warriors in "Hôichi the Earless." No one is there to see her, but she is seen, from inside. Yumiko walks to the window and peers in. The camera watches her from outside and behind. First an empty reaction shot of Yumiko, taken from the phantom perspective of an absent Ikuo, then her view into the empty factory space where she had earlier observed him. No one is there now. But no one was ever there. Ikuo's blank stare in response to Yumiko's exaggerated expressions in the earlier episode suggests that he was already absent then, already translucent and phantasmatic. (In the subsequent scene she teases him for not having noticed her watching him. "How could I know?" he says. "*I* would have," she answers.)

In the hallway of the apartment she once shared with Ikuo, Yumiko walks away from the camera, toward a hazy light cast by a semiopaque window. As she reaches the entrance to her apartment, she stops and turns. She is framed in a low-angle long shot. A cut moves the camera closer and frames her in a medium shot. Slowly Yumiko turns her head toward the right and, for a brief instant, stares into the camera, which seems to have taken the place of some absent figure. An arrest, momentary and photographic, lost in the resumption of time. Ikuo is in the camera, is himself the camera. An unseen energy. Slowly Yumiko pulls away from the eyeline match, breaking the phantasmatic contact with the empty figure there. Her gesture is transposed from an apparent perception (of a phantom figure located at the point of the camera) to a contemplation. There and not there, in an instant. A flash.

Ikuo is in fact there and not there, always. Never fully there when alive, he is never entirely absent after his death. His trace is carried, most materially, in the various manifestations of the bicycle. In a Notô marketplace, Ikuo's son, Yûichi, shows interest in a green bicycle as Yumiko watches. She is struck it seems, by Yûichi's choice, which echoes the bicycle his father left behind. (In the opening scenes of *Maborosi,* Ikuo, who has stolen a bicycle, paints it green.)[3] *Maborosi* is haunted by bicycles, by one bicycle in particular, which serves, perhaps, as a metonymy of the trains that map the world

Yumiko, in Kore-eda Hirokazu, *Maborosi* (*Maboroshi no hikari*, 1995).

of *Maborosi*. (In numerous scenes, bicycles and their riders are set against trains that speed by in the background.) One night, after returning home drunk to a worried Yumiko, Tamio hears the bicycle key in Yumiko's hand. The sound resembles the bells that echo throughout the film. He cannot see her; his head resting on the table, turned away. Although drunk, he hears the sound of the bell in Yumiko's hand and asks what it is. "What did you just hide?" Yumiko insists that she wasn't hiding anything; she had planned to remove the keychain bell from an old bicycle key. The phantom bicycle, the sound of this transportation that sutures spaces and histories, bodies and generations, remains the only audible trace of an otherwise silent Ikuo. He is constituted aurally and in the dark by this sound, a metonymy of the bicycle he rides, leaves behind, and has become.

The confusion of reality with dream spaces, a trope that echoes Mizo-guchi Kenji's *Ugetsu* and Kobayashi Masaki's *Kwaidan,* adds to the obscu-rity of *Maborosi,* a shadow world suspended between life and death, past and present, visibility and invisibility, visuality and avisuality. From her window early in the morning, Yumiko watches Tomeno, a local fisherwoman, leave for the ocean. The dark room is pierced only by the small opening of the window from which Yumiko peers. The dark morning provides a deep blue light to the otherwise black space of the room, which fills most of the screen. Like a dreamscape that opens outward and beyond from the camera obscura in which Tamio sleeps, the scene of Tomeno's departure like a projection. Tomeno notices Yumiko watching and the two converse; Tomeno speaks, Yumiko responds with hand gestures. In a cut, the camera leaps out of the room, outdoors, into the space where Tomeno stands. Tomeno moves away from the camera, away from the place where Yumiko watches, and leaves the frame to the right. Tomeno's movements echo those of Yumiko's grandmother in the dream-memory that opens the film. A haunted gesture. A cut returns the view to Yumiko, seen from the side, the profile of her face barely visible in the faint twilight. The scene ends when Tamio calls to Yumiko from the dark, from offscreen, "What are you looking at?" Yumiko turns away, leaving a dark frame with a small sliver of dark blue.

Yumiko closes the window and returns to bed. When Tamio asks if she is homesick, if she wants to return to Osaka (Amagasaki), Yumiko says no. "I just had a dream. . . . it woke me up." A dream or the scene of Tomeno, an old woman like Yumiko's lost grandmother, leaving into the dark? The interior and exterior worlds are confused in *Maborosi,* fused together in an uneasy balance—an emulsion marked by the shifting lights that illuminate the film.

The uneven light of *Maborosi,* an effect of the natural underlighting that Kore-eda maintains throughout the film, establishes a dynamic and pulsating luminosity. The many scenes of people and vehicles passing through tunnels create opportunities for the lighting to change from dark to light and vice versa during a single shot. Yumiko's face against the window of the police car that takes her through the rain to Ikuo's lifeless body, her face in the train as she watches Yûichi and Tomoko beginning to form a friendship, pass from light to shadow, visibility to obscurity. The rain that falls on the police car window is like displaced tears; her emotions have moved outside her body and into the environment. The interiority that fails to erupt at this moment of her loss returns in the form of an exterior element, rain. Her interiority takes place outside and returns to her like black rain, poisoning and marking her from the outside. The car window can be seen as a displaced surface, a second skin. (All of the planes in *Maborosi* can be seen perhaps as displaced human surfaces, the film itself a translucent body.) Light pulses and pulsates in *Maborosi,* adding a type of liminal vitality to the film, a phantom light. Although a dark film, a film about darkness, nearly invisible and impenetrable, *Maborosi* establishes a

Yumiko's displaced tears, *Maborosi.*

dark luminosity. It follows the logic of Tanizaki's shadow—a darkness that is posited, illuminated in the dark by other darknesses. Darkness, a form of phantom light.

In the film's closing scenes, Yumiko, who has perhaps decided to flee Notô, sits inside a small structure that serves as a bus stop. An attempt to leave like her grandmother, like Ikuo; to move to the other side. She is barely visible, a dark figure inside a dark space. Seen from a distance, a bus arrives and leaves, Yumiko's decision not yet apparent. Slowly Yumiko emerges from the structure. She notices a funeral procession as it moves toward the ocean. In an extreme long shot, the procession moves horizontally across the frame, set against the dark sky and ocean. As the procession moves from right to left, Yumiko enters the frame, following the procession from several paces behind. Tamio arrives in a car, drawn to the ocean's edge by the funeral pyre. The funeral is over, the fire continues to burn. Still in extreme long shot, Tamio and Yumiko face each other. As she moves

Maborosi.

toward him from left to right, he retreats, moving backward at first, then turning around and walking to the right of the shot. From a distance, their intentions are hard to discern. The camera follows their movements in a pan. Two dark figures, silhouettes, against the purple sky and dark ocean. "I just don't understand," Yumiko says, "why he killed himself. Why was he walking along the tracks?" "Why do you think he did it?" she asks Tamio. Tamio answers after a pause: "The ocean calls to you, he said. My father used to go out to sea. He says when he was out alone, he would see a beautiful light, shimmering in the distance, calling to him. I think it can happen to anyone." The final scene invokes the first. Despite Yumiko and Tamio's visibility in an extreme long shot, their voices are acousmatic, disembodied, phantasmatic. Against the sky and sea, Yumiko and Tamio are distant, atomic figures. Reduced to atoms, swallowed by the elements, by the post-catastrophic world. Only their voices remain, it seems, only their voices are visible.

In the end, Ikuo's interiority, the inaccessible forces that drove him to take his own life, may have come from outside. But from a displaced, phantasmatic outside. The phantom light, the seductive phantasm that calls when one is alone, serves as the source of Ikuo's desire. In his case, the light of the oncoming train, perhaps. Ikuo is reduced to a light, has become himself a solitary light, a projection dispersed atomically into the world.

Body and voice, life and death, inside and out, visual and avisual light, exist in *Maborosi* as suspended dialectics. As emulsions. Yumiko's search for an explanation of Ikuo's death, for some insight or glimpse inside that might reveal his motivation, his desire to die, conscious or unconscious, is impossible in a world where the lines between inside and out have been obscured. A world inside out, depth rendered entirely on the surface, visibility marked by a profound and consistent invisibility. What is lost with Ikuo is not only his psyche and its secret archives but the psyche as such. *Maborosi* can be seen in this sense as a profoundly nonpsychological film, apsychological, without psyche, a psyche transposed without, on the outside. An exteriorized, superficial psyche that cannot be probed because its depths are on the surface. Ikuo, alive and dead, represents the film's displaced interior. He is inside out, an inside projected outward, visible only on the outside as an aftereffect of his life. When visible, his face is inscrutable; his actions impenetrable, his figure entirely avisual, Ikuo is an invisible man, radiating a dark luminosity over the world that surrounds him. Like Tanizaki's Oriental body, Ikuo projects a darkness outward that stains the world around him; he casts a shadow over the film, extends his darkness across it, rendering the film, like him, opaque and obscure. This darkness is Ikuo's interiority, his impenetrable psyche, dispersed atomically across the film's surface. It is visible in the first instance only from the outside. Even more than his visage or figure, his identity or interior, Ikuo's darkness comes to saturate the film and its world. He is a skiagraph, a phantom and illusion, a *maboroshi*. An atomic light projected into and out of a shadow world renders the world blind, blinded like angels. A shadow optics.

• • •

One hundred years after H. G. Wells's 1897 *The Invisible Man,* Kurosawa Kiyoshi's *Cure* portrays another type of invisible figure in the person of his protagonist, Mamiya Kunihiko. Mamiya, an amnesic former psychology student, hypnotizes his victims, who in turn kill others, often those close by. In each instance, the mesmerized killer is able to remember the murder, its precise details, but not its rationale or motivation. None are able to recall their encounters with Mamiya. It is as if he vanishes on the occasion of the encounter, entering, like a phantom, their exteriorized psyches, but leaving no trace.

    Cure operates in at least two genres, psychological thriller and horror, and follows the protocols of each: the logic of detection and the force of supernaturalism, causality and catastrophe. The two worlds are never reconciled, one order never supersedes the other. *Cure* remains in an uneasy balance between the two, between increasing clarity and deeper obscurity. Between a movement toward the surface and an abyssal plunge. Since Mamiya does not himself commit murder but acts rather as an instigator, the very structure of crime is displaced in *Cure*: the crimes occur elsewhere, as deferred aftereffects, outside the economy of criminals and victims. As the film unfolds, Mamiya's method gradually becomes visible: he hypnotizes his victims using the flame of a lighter or running water, forcing each victim to confront some aspect of his or her own psyche from the outside. Each killing is marked by an "x" carved on the victim's body. A signature of sorts, an operation that opens the body of the other surgically, and also an erasure. *Cure* carries throughout its narrative the complex semiology and asemiology of the figure "x." Sign and design, an ex-sign.

The film opens at the scene of a murder. A man has killed a prostitute, leaving an "x" on her corpse. Mamiya is first seen at the beach, where he meets Hanaoka, who brings Mamiya home and tries to solve the riddle of Mamiya's amnesia. Hanaoka discovers Mamiya's name inscribed on his clothes, but Mamiya has lost his memory and has no sense of his own identity. He seems to forget everything from moment to moment, often repeating the same questions shortly after they have been answered. The next morning Hanaoka tries to kill himself after realizing that he has murdered his wife. After Hanaoka, Mamiya's next victim is a policeman, Oida, who brings Mamiya to his police box after watching Mamiya jump from a small building. During the night, Mamiya places a lit lighter on the table and says to Oida, "Look at this. You hear my voice, don't you?" The next morning Oida kills his partner by shooting him in the back of the head for no apparent reason while they prepare for their morning rounds. Mamiya moves, it seems, aimlessly from victim to victim, each new victim the result of a chance encounter.

Takabe Kenichi, the lead detective investigating the series of apparently related but seemingly inexplicable murders, is warned by his friend and consultant, Sakuma Makoto, a psychologist, not to "get too deep [*fukairi*] inside a person's soul." Like the deaths of *Maborosi,* the grisly murders in *Cure* seem unmotivated. They fall outside the realm of psychology, resisting the idea of a deeply interiorized cause. In each case, the perpetrator admits the crime, but is unable to provide a satisfactory motivation for it.

Mamiya has no inside, no interiority, a condition manifested in the film as amnesia and the absence of self-knowledge. He explains his condition to Dr. Miyajima, the woman doctor whom he also entices to kill. When Miyajima asks Mamiya during a medical examination if he has any worries, any unease, he responds with a provocation, "You're the one with worries." When she presses Mamiya on his assertion, he claims to have forgotten. Running water from a tap, Mamiya begins to fill a glass of water while engaging the doctor in conversation: "All the things that used to be inside me are now on the outside. So I can see all of the things inside you, doctor. But I myself am empty." He is apparently aware of his emptiness, which he links to an enhanced ability to see others, their interiorities in particular. One precipitates the other, an exchange of one's own interiority for extrasensory vision.

Mamiya knocks the full glass of water over, causing the water to spread across the floor.

In Mamiya's brief discourse, the first sign of self-knowledge, he renders himself inside out. His interiority has been transferred outward into the world, *expressed*. He can now see the insides of others, but is himself empty. His emptiness induces a form of X-ray vision, and his invisibility is an effect of his emptiness as is his extravisuality, x-visuality. Mamiya is a type of invisible man; he sees because he is no longer there but elsewhere, outside, everywhere.[4] Mamiya returns memories and desires (and later future images) to his victims from the outside; he restores to his victims an unfamiliar interiority, like an X-ray image. An outside interiority. Mamiya himself leaves no impression on those he comes into contact with; he is unremarkable. No trace or memory of an encounter remains.

An antecedent to Mamiya can be found in Dr. James Xavier, the protagonist of Roger Corman's 1963 *X: The Man with the X-Ray Eyes*. Xavier develops a method "to sensitize the human eye so it sees radiation, up to and including the gamma rays and the meson wave." He becomes his own first test subject. The first effect on Xavier's vision after placing drops of the serum into his eyes—they serve as synthetic or prosthetic tears, perhaps, like those of Niki Junpei in *The Woman in the Dunes*—is that he sees through his own eyelids. He becomes perpetually vigilant, unable to stop seeing. "It's like a splitting of the world," he says, invoking the nuclear age, a premonition of total catastrophe destined to follow. "Vision is fragmented," he says in disjointed syntax, "more light than I've ever seen." Xavier's vision is atomic, a vision of atomic apocalypse, but also an atomic vision that annihilates the world in his look. Xavier's new eyes give him penetrating X-ray vision, a fantasy Xavier shares with Freud, at the expense of his sanity. The effects of X-ray vision are irreversible, and they continue to develop, penetrating further and deeper until the world is rendered transparent, limitless, and empty. By penetrating every surface to the ends of vision, nothing remains hidden. Nothing remains unexposed. Like Tanizaki's nightmare of a world overilluminated, "the man with x-ray eyes" has annihilated darkness, transforming the world into a luminous void, an abyss, nothing. Like all of the versions of invisible men before him, Xavier is driven mad by his excessive visuality.

At first, Xavier uses his enhanced vision for medical purposes, diagnosing a girl's condition by seeing through her body to the source of her ailment. An organic X-ray machine. Then, he uses his talent for pleasure, seeing through the clothes of young women at a party. The shift from science to pornography underscores the unstable contours that shaped the X-ray image from the moment of its appearance. After inadvertently killing a colleague, Xavier slips into the underworld, serving as a carnival attraction, "Mr. Mentalo," then as a "healer," a hustler spiritualist. All the while, he seeks to reverse the effects of his X-ray vision.

Xavier's world has become completely illuminated, ecstatic. Absolute and excessive in its clarity. "I'd give anything to have dark," he says. While attempting to flee his captive existence, Xavier comments on the world he sees around him, the world he creates with his monstrous vision. "A city unborn. Its flesh dissolved in an acid of light. A city of the dead." Excess light destroys the world, dissolves it, and turns the world into a "city of the dead." A postatomic world. To continue to fund his research, Xavier travels to Las Vegas, where he uses his extrasensory perception to gamble. He sees through slot machines and through cards, winning at every turn. Ultimately, Xavier raises too many suspicions. Confronted by the casino management, he is unmasked, his eyes exposed. His eyes have changed physically: they have darkened, gold pupils surrounded by black, seared, perhaps from the excess and constant force of light. Eyes "melted out of sheer ecstasy," to return again to Willem de Kooning's expression.

Xavier has become blinded by his extravision, his eyes burned in his face. "So blinded by ambition," reads the promotional material for *X: The Man with X-Ray Eyes*, "that he dared to glimpse eternity." X-ray vision and blindness, visuality and time: the lines between visuality and visibility, visuality and time blurred into an apocalyptic avisuality. Xavier flees the casino and drives into the desert, pursued by the police. He wanders deeper into the desert, toward the end of the world, toward eternity. The film's last scene takes place in a prayer meeting. The preacher asks Xavier, who walks into the revival, "Do you wish to be saved?" "Saved?" he responds, his eyes now completely black, "No, I've come to tell you what I see. There are great darknesses. Farther than time itself. And beyond the darkness, a light that glows and changes. And in the center of the universe, the eye that sees us all . . ." The preacher responds, "You see sin and the devil. But the Lord has told us what to do about it. Said Matthew in Chapter Five, 'If thine eye offends thee, pluck it out!' Pluck it out! Pluck it out!" As if realizing the solution for the first time, Xavier tears his eyes from his head. The final shot of the film is of Xavier's empty eye sockets.

Xavier's X-ray vision hollows out the world, flooding it with light until every trace of darkness is dispelled and the world is exposed to the searing light of atoms. The archive has been burned, leaving behind a vast emptiness. The emptiness returns to Xavier's eyes, which, like those of Brother Jack, the nemesis of Ellison's invisible man, suffer from a fragile relationship to his body. They have been swallowed by the emptiness, by the unseen energies of the world that render Xavier, in the end, empty. And like Xavier,

Mamiya survives the emptiness that he witnesses and produces by becoming the emptiness himself, by merging with it, entering in and out of the emptiness through the sign "x." In *Cure,* the "x" sign is a mark of erasure, an allusion to X-rays, perhaps, to X-ray vision, but also an opening: the slices through which interiority escapes to the outside, and the exteriority enters inside. It leaves both sides of the world exhausted, depleted, and empty.

Kurosawa describes his desire to follow the originary emptiness he portrays in *Cure* by selecting locations that illustrate the emptiness of his characters. In his 2001 collection of essays, *Cinema Is Horrible (Eiga wa osoroshi),* Kurosawa explains that he sought in *Cure* to portray the sense that each character is inflected by some form of emptiness, carries somewhere in her or in him an irreducible emptiness. He says: "Of course a person's soul cannot be photographed. So I decided to choose as my locations spaces that felt somewhat empty.... But how can one photograph emptiness?"[5] Interiority projected onto the world, the world as a vast and displaced interiority. The invisible inside of each person, the empty interior that haunts each character, explodes into the world as a material interiority thrust outward. What is visible in the world as the world is an emptiness made visible on the occasion of its projection.

At the scene of her hypnosis, a displaced, inverse seduction, Miyajima seems lulled, entranced. After several cuts between Miyajima's subdued face and the streaming water, she looks toward Mamiya, who responds, "Don't look at me." She looks away, down toward the floor. She is absorbed, in the sense that Michael Fried defines absorption, absorbed in herself, but is also absorbing herself from the outside, a self reintroduced to her by Mamiya. "Now tell me about yourself," Mamiya says. "You're just a woman. Why did you become a doctor?" "Just a woman?" Miyajima asks. To which Mamiya responds, "That's what people said, isn't it?" Mamiya places his hand on Miyajima's head and relates to her a memory from her time as a medical student. He describes her first dissection of a human corpse, a man. He tells her it was the first time she'd seen a man unclothed, that it felt good to cut into his flesh with a scalpel. Mamiya's memory from without ends when he tells her that what she really wanted was to cut men open. Mamiya throws a glass of water on Miyajima's face, which brings her violently back to consciousness. After Mamiya leaves, Miyajima discovers a large black dripping "x" painted on her wall. She wipes the "x" from the wall as the scene ends.

During Oida's interrogation, Sakuma reproduces Oida's original hypnotic state, inducing him to perform the gesture of carving an "x" into a detective's chest. The scene cuts to a subjective shot, handheld. The camera moves around a corner and into a men's room. As the lavatory comes into

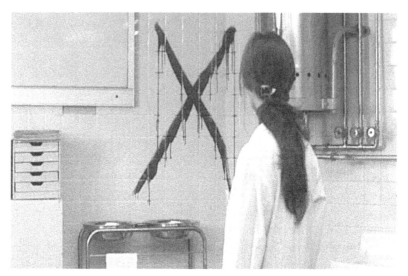

"X," Kurosawa Kiyoshi, *Cure* (1997).

view, a man is seen on the ground, bleeding heavily, with Miyajima crouched over him holding a scalpel to his neck. The camera stabilizes, then cuts 180 degrees to a view of the scene from the other side. No one holds the point of view of the previous shot, no subject fills the space of the subjective shot. A man walks into the men's room and stops when he sees the scene. From his point of view, Miyajima tears the face from her victim. She looks up at him, and the eyeline match confirms it as a reaction shot. Miyajima looks confused as she proceeds to remove the skin from her victim's face.

Meanwhile, Takabe's investigation continues. He locates Mamiya's residence, a small room, uninhabited, he learns, for the past six months. In Mamiya's room, Takabe discovers books on Franz Anton Mesmer, psychological and psychoanalytic theory, suicide, and depression, among others, and a paper by Mamiya on animal magnetism. Mamiya's research interests appear to include theoretical and technical aspects of psychology, its history and practice. Mamiya's focus appears to be on the vicissitudes of identity, in particular, the movement of identity from one being to another. A phantasmatic movement between beings. Mesmer, pronounced in Japanese *mesumâ,* rhymes with Mamiya.

Like *Maborosi, Cure*'s composition includes numerous single lights, small luminescent spots against and from within dark screens. They form small moments of avisuality, an immiscible mixture of invisibility and hypervisibility, like the "thermoptic" invisibility described in Oshii Mamoru's 1995 *Ghost in the Shell (Kôkaku kidôtai),* a manipulation of heat that reduces the temperatures of visibility, rendering one invisible. Flickering neon and fluorescent lights in dark enclosed spaces, blinking city lights in the distances, the small flames and dripping water that mark Mamiya's seduction. But even diffuse light can produce the effect of a small, pointed light. When Takabe decides to incarcerate his wife, Fumie, who has been throughout the film slipping into psychosis, the two are seen seated in the rear of a bus. Behind them and through the rear window, clouds envelop the bus as if they are in the sky, flying. Takabe and Fumie have been swallowed by the bright sky, by the vast aerial expanse that seems to absorb them. A projection, perhaps of Fumie's interiority, or of Takabe's. Lost, light, floating. The scene returns later.

Eventually Mamiya is apprehended, or rather ceases to resist. He has never really sought to evade the authorities, wandering carelessly in and out of their grasp. The conclusion of *Cure* represents something of a narrative collapse: the slow development of the story shifts to an accelerated collection of flash scenes in which the distinction between reality and hallucination vanishes. A series of images erupts inside Takabe, leading him to hallucinate the scene of his wife's suicide. Unable to resist a sudden urge, Takabe rushes home to discover Fumie hanging from a rope in the kitchen. Takabe falls to the floor screaming. His fantasy, projection, wish is broken by Fumie, who is suddenly in front of him, alive. She asks him, "What's wrong?" Like dreams, these scenes are marked by their urgency, their sense or impression of reality. And by wish fulfillment. Visually, they are indistinguishable from any other scene, framed as fantasies or illusions only in transition. The criminal case gradually recedes, taken over by a series of unstable scenes and elements no longer bound within the logic of detection.

Near the end of the film, Sakuma shows Takabe a new piece of evidence, a primal scene of sorts, of cinema and hypnosis: an 1898 film that features a woman hysteric, Murakawa Suzu, who, Sakuma says, killed her son and cut a cross in his neck. Sakuma dates the work from the late nineteenth century, adding that it represents the oldest record of hypnotism in Japan. In the brief *actualité*, a hand barely visible at the edges of the frame gestures in the shape of an "x." A hypnotist, perhaps, a radiographer who signs "x." Sakuma speculates that Mamiya follows a long line of hypnotists, spiritualists, and occultists who have been suppressed politically throughout the modern history of Japan. Sakuma sits in a chair, his face and body darkened by a shadow. The camera moves toward his face. Abruptly, the scene cuts to Mamiya's apartment, furnaces blasting, animals in cages. The camera seems to single out a caged monkey. The subjective camera moves through Mamiya's residence, which detectives are searching for evidence. For the first time, Sakuma appears in the scene. He walks to Mamiya's desk and lifts a book titled *Heresies* and opens it to chapter 4, "Mesmerian." Sakuma turns the page to the image of a man, his face erased, whose birthdate is filled by a question mark, his death inscribed as 1898. The scene changes. Sakuma stands outside a large, dilapidated building. Someone barely visible, perhaps the faceless man in the image, peers at him from behind a window. Inside the building, apparently the psychiatric ward where Mamiya is held, Sakuma enters Mamiya's empty cell room. The camera pans around the empty room, then moves along with Sakuma toward a source of light. In the bath area, an animal, a small mummified monkey, has been strung above the bathtub in a pose suggesting crucifixion. The camera swivels swiftly to a dark corner of the room, followed by a shot of Sakuma repeating the movement, turning his head toward the dark corner. A figure emerges from the darkness, it is Takabe. He moves toward the camera and, once the shot changes, toward Sakuma, who retreats into a corner. As Takabe presses toward Sakuma, a voice, "Sakuma ... Sakuma ..." The scene returns to Sakuma's apartment. He is seated in a chair in the corner, his forehead drenched in sweat. The previous scenes have been a hallucination, formed perhaps from a mixture of memory and fantasy.

"Sakuma," Takabe continues, "so what is Mamiya? What's your guess. Tell me." "A missionary [*dendôshi*]. Sent to propagate the ceremony." Suddenly, the spell seems broken. Sakuma laughs, "I'm imagining things. What am I saying? Takabe, don't take me too seriously. I'm a little tired. Let's call it a day. Now *I'm* getting in too deep [*fukairi*]."[6] Sakuma enters his bedroom

"X," *Cure.*

Phantom Cures

150

and turns on the light, revealing a large "x" marked on his wall. Takabe notices the mark and asks about it. Sakuma seems confused by the mark as he frantically tries to erase it from his wall. "You saw Mamiya," says Takabe. Sakuma denies it, but Takabe insists. "It's strange," says Sakuma, "but I can't remember."

In the next scene, Mamiya escapes from the ward where he is held. Takabe receives a phone call in his car. Takabe returns to Sakuma's apartment and enters the scene of an investigation. He learns that Sakuma has killed himself. Takabe listens impassively to the policeman's account; the room is splattered with blood. Takabe is seen again seated in a bus, the sky opening behind him, then walking into the large hospital seen earlier in Sakuma's fantasy. Inside one of the dark rooms, Takabe finds a photograph of the faceless boy, suspended in the air.

Kurosawa describes his attempts to represent the avisuality of fear in a material way. He adopts the idiom of space and geometry, surface and interior. "Fear (= death) is everywhere on the screen. At the same time human beings (= life) suddenly emerge from inside. Only cinema, perhaps, can manage such nonsensical trick representations."[7] Only cinema can cover the screen with fear (itaru tokoro ni kyôfu wo haritsukeru) and frame inside this metaphysical surface a human figure.

Takabe enters a semienclosed room and sits on a bench. The camera stays on him for several seconds, unstable and trembling slightly, before panning to the right, toward a doorway through which Mamiya enters. "So you've finally come, detective." The camera follows Mamiya until he reaches Takabe, still seated, holding his head in his hands. The two are framed together in a manner that evokes the scenes of Mamiya's assaults. "Why did you let me escape?...I know why. You let me go so that you could learn my true secret...all by yourself." The handheld camera tracks Mamiya as he walks away from Takabe and toward the camera. A gulf opens between the two. Mamiya stops and sits down. In the distance, Takabe still sits with his head lowered. Mamiya continues. "You didn't have to do that. Anyone who wants to encounter his or her own true self always comes here. It's fate." Takabe rises abruptly and, reaching into his breast pocket while walking toward Mamiya, draws a revolver from his pocket and shoots Mamiya, who collapses after the first few shots. Mamiya falls out of the frame, Takabe continues to advance. A reverse shot shows Mamiya on the ground, bleeding, Takabe standing over him. A low angle shot of Takabe in close-up. "Do you remember now? Do you remember everything?" Mamiya nods. "You do? This is the end for you." Mamiya raises his hand in the air, his index finger pointing toward Takabe. Mamiya's hand is shown in close-up as he begins to sketch an "x" figure.[8] As he completes the gesture, Takabe's

"X," hand and finger, *Cure*.

gun enters the frame from the right, counteracting the line of Mamiya's hand, and begins to shoot. In a long shot, Takabe continues to fire into Mamiya's body.

Takabe enters a flooded examination room, the same one apparently featured in the films of Murakawa Suzu. He finds an antique phonograph, sets the needle on the disk, and sits down to listen. A man's voice utters a series of brief, mystical, and perhaps therapeutic statements that refer to healing and existence. Perhaps a hypnotic incantation. The scene returns to the other hospital, where Takabe's wife, Fumie, resides. A nurse walks down a dark corridor toward the camera, a scene now familiar as a refrain in *Cure*. She stops and turns toward a sound that rises from behind her. For a brief moment, an image of Fumie appears, only her head, apparently on some type of moving crucifix. A quick cut to Takabe alone, eating dinner in a restaurant. He is well groomed and seems invigorated, his plate empty. In a previous scene at the same restaurant, he had barely touched his meal. He pushes away his plate, asks for coffee, and reaches for a cigarette when his mobile phone rings. He responds to the call buoyantly, unaffected: "Yes? Alright. Bring the car around." He puts down the phone and lights a cigarette. Takabe seems relaxed. The waitress brings his coffee. A close-up of Takabe's face in profile. He continues to smoke as the waitress clears his place and leaves. The camera wanders away from Takabe and, refocusing, finds the waitress in the distance. It tracks her movements from a distance, although it is not clear if the views of her are from Takabe's perspective. A woman, perhaps her supervisor, approaches her from behind and, placing her hand on the waitress's shoulder, says something to her.

▼

152

The waitress nods, and as her supervisor walks away, she seems briefly absorbed in thought. The waitress resumes her work, and the camera again follows her. The point of view is close to where Takabe is seated, but remains ambiguous. As the scene and film draw to a close, the waitress picks up a butcher knife and, still seen in a long shot, walks to the right. At the very limit of the film, at its virtual end, the cycle seems to have resumed, triggered perhaps by Takabe, who is now Mamiya, or some version of him. The scene cuts to a view of a dark street that recedes into the distance. Credits appear on the screen through what appears to be a shattered glass surface, a window, which was not visible until the lettering appeared. The names of the cast are fragmented and lacerated. They form an invisible, shattered surface.

The final moments of *Cure* are not entirely legible, the conclusion elusive. Takabe has been revitalized; he receives news of his wife's death, or perhaps of another death. The strange ellipsis of perspective and the focus on the otherwise marginal figure of the waitress suggests perhaps the dissolution of Takabe as a single subject. He may be inhabited or possessed by another, maybe even by Mamiya, but he is no longer haunted by a desire he cannot express; he has been cured of his desire, which has been released. Takabe has been released from the shadow that has emanated from him throughout the film; Mamiya has been released into the world. If Takabe is now Mamiya, if Takabe has incorporated Mamiya, then Mamiya, the man with no interiority, has himself become an interior. Mamiya has filled Takabe with (his) emptiness, with the emptiness that he is. An interiority constituted by the lack of interiority. Takabe is filled with emptiness. By entering Takabe, Mamiya has entered the world: he has been dispersed atomically into the outside.

Mamiya has inhabited Takabe, like the sun god Aten in Freud's speculative account of Mosaic Judaism. "The central fact of the development of the Jewish religion," says Freud, "was that in the course of time the god Yahweh lost his own characteristics and grew more and more to resemble the old god of Moses, the Aten."[9] This incorporation keeps the other alive and active, one *body* hidden within another. Like Takabe and Mamiya, Genjuro and Wakasa, Hôichi and the Heike phantoms, like atomic radiation: an interiority haunted and constituted by an immaterial exteriority. Mamiya can be read against the screen of Freud's speculation on Moses, a "missionary [*dendôshi*]," as Sakuma calls him, "sent to propagate the ceremony."[10] Mamiya's transmission of the ceremony from one body to another, one culture to another, one epoch to another, takes the form of a cut, an incision that marks the body. X. One could say, a circumcision. Circumcision, the originary violence that founds the Jewish community and marks its members as such was, insists Freud, an Egyptian custom, whose origins were erased. An "immemorial archive" right on the body and exterior, says Jacques Derrida.[11] The laceration passes from one body to another, its source erased. Mamiya's modus operandi, MO, his Moses.

Among the effects of a hidden religion, a secret God that takes up residence in the corpus of another, is a figural invisibility or, more precisely, avisuality.

> Among the precepts of the Moses religion there is one that is of greater importance than appears to begin with. This is the prohibition against making an image of God—the compulsion to worship a God whom one cannot see. . . . But if this prohibition were accepted, it must have a profound effect. For it meant that a sensory perception was given second place to what may be called an abstract idea—a triumph of intellectuality over sensuality or, an instinctual renunciation, with all its necessary psychological consequences.[12]

Intellectuality over sensuality. The very dilemma of *Cure,* resolved in the form of an instinctual renunciation: Takabe renounces his desire by inviting the other inside, by inviting the emptiness of the other inside him. By taking on—incorporating—the desire of the other in place of one's own. "The fantasy of incorporation," Nicolas Abraham and Maria Torok say, "simulates profound psychic transformation through magic."[13] A magical transformation of the psyche, achieved through a magical contact and alliance with the lost other. "So in order not to have to have to 'swallow' a loss, we fantasize swallowing (or having swallowed) that which was lost, as if it were some kind of thing."[14] The fantasy of swallowing an other through an orifice, an opening, or cut. This becomes the solution or *cure.* "The magical 'cure' by incorporation exempts the subject from the painful process of reorganization."[15]

Another feature that remains obscure in Kurosawa's film is its title, *Cure,* which like the film's protagonist represents an avisual dilemma: visuality without visibility. The title, offered in English and without a definite article, hovers between a noun and a verb, a solution and an imperative. The cure or to cure. And of what? What is the cure imagined by this film, what is the illness or ailment? Suggested throughout the film is Takabe's affliction, Takabe as the truer source of disorder, of disease, a condition reinforced by Mamiya's practice, which reintroduces from the outside the internal anxieties and disquiet of others. The doctor treating Takabe's wife says to him, "From my perspective, you look sicker than your wife." The cure, as an antidote and command, can refer to the specific needs that are expressed by the victim/murderers in their crimes, but also to a structural restoration of interiority from the one place where it can be seen, the outside. In and by another, an other that sees what the subject cannot see.

Obscure. A psychoanalysis, a psychoptics. A spectral visuality, X-ray psycho-analysis, X-analysis, from the outside in, ex-analysis.[16]

Takabe suffers, it seems, as do all the characters of *Cure,* from an excess of self, from a desire that is forced outside. Or rather, from a self consti-tuted only as an excess, accessible only outside as an effect of the other. Each character suffers from the irreducible exteriority of oneself, the impossibil-ity of being one, and the shock of discovering oneself outside. The cure arrives in the destruction of another, of the other that sustains the surplus. But not only in destruction, as the ending suggests, but rather in yielding to the outside, to the other from the outside. To be cured (of oneself, of a desire that is always outside) is to open oneself to another, to allow another to enter from the outside. No longer oneself, oneself alone, the archive arrives from the other, is itself other. The other archive.

$$\bullet \quad \bullet \quad \bullet$$

Maborosi and *Cure* portray dark worlds forged in the collapse of light and psyche. The inside erupts outward, filling the atmosphere with interiority; the outside pours inward, suffusing the inside with the world. Between obscurity and emptiness, interiority and exteriority, deep and flat space, *Maborosi* and *Cure* stage a global fragmentation of surfaces that turns the inside out, the outside in. Interiority swallowed by the unseen energies of the world, psychic emptiness pouring outward like invisible ink, an inte-rior and avisual black rain and light that stain the atmosphere, rendering the outside unconscious. In both films, the world is no longer one, no longer contained, anarchived. The worlds are porous, a series of atomic surfaces that allow the light to pass through each limit to the other side. As each threshold is passed, another world opens up beyond the surface. Another side. Each film maps a secret exit, a way out into and through the surface, to its other side, the other side of cinema. Each film vanishes into the surface, leaving only a surface, obscure and empty, a flash of atomic light: atomized and destructive. The darkening world of *Maborosi* dissolves into the material emptiness of *Cure.* Obscurity and emptiness, two forms of avisualizing space, of establishing the material avisuality of a cinema, are no longer visible in these two films, which establish two limits of a Japa-nese visual economy at the end of the century still haunted by the psychic, aesthetic, and violent experiences of World War II. A world swallowed by darkness, a universal darkness. A dark archive formed in the shadow, and by a shadow optics. By a secret and other archive, an other mode of seeing, anarchiving.

Each film establishes an address to an other, to an other that withdraws; trace and phantom, an other that refuses to step into the light, casting the avisual shadow of an atomic sensuality, a shadow optics. An other dispersed throughout the atmosphere, everywhere, in everything, like Tanizaki's shadows, and nowhere. An atomic other. *Maborosi* and *Cure,* phantom and cure, chart secret passages to another, to a secret other, a secret you. *Maborosi*'s Yumiko seeks a response from the figures that abandon her, the phantoms that slip into the darkness—into night, into dreams—and fill the world with their darkness, traces of their disappearance. Her world is filled with a dark disappearance. The response comes to her in the form of a phantom light, a single light at the edge of the ocean. Takabe seeks a cure, a respite from the madness that seeps into the world around him and overtakes him. A madness that reaches him through others, through Fumie, Sakuma, Mamiya. A secret orifice, like the illicit puncture that Freud dreams in Irma's body, through which the madness enters him. Takabe is able to achieve a measure of relief or evacuation through his contact with Mamiya, who cures Takabe of himself. Mamiya provides Takabe with emptiness, fills him with it, allowing him to experience the voluptuous emptiness of an other, of otherness itself, as a deep and material vacuum.

An encounter with nothing at the limit, at the end of light and of cinema. An end already figured in the "unseen energies" apparent in its beginnings. Swallowing space. Paul Virilio imagines a "sightless vision" at the end of cinema. The spectator replaced by a "vision machine": a machine that controls a camera, a camera that returns "images" legible to the machine, bypassing the "televiewer" entirely.[17] The virtual image comes, for Virilio, at the end of cinema, at the end of a system of viewing relations no longer necessary and no longer possible. "Every image (visual, sound) is the manifestation of an energy, of an unrecognized power."[18] One effect of the vision machine, for Virilio, is a transformation of images into energy, into a "sightless vision" and ecstatic blindness.

> Blindness is thus very much at the heart of the coming "vision machine." The production of *sightless vision* is itself merely the reproduction of an intense blindness that will become the latest and last form of industrialisation: *the industrialisation of the non-gaze.*[19]

The vision machine arrives at the end of cinema, marks the end of cinema, and ends cinema as a visual phenomenon, converting—or restoring—it to a set of avisual energies.

The technocentric future that Virilio imagines, the transformation of visuality into vision machines that engender sightless vision, invisible and avisual images, was already at work in the eruption of radical interiority in 1895: psychic, corporeal, and vital interiority were already marked as avisual and mediated by vision machines—psychoanalysis, X-rays, and cinema. These vision machines, apparatuses, techniques, and technologies were already dismantling the visible world, producing an irreversible *démontage* of the world of images and the image of the world at the fin de siècle.[20] The transposition of interiority into the world, its expression, was already the moment of a lost visuality.[21] Cinema has always been a vision machine, a secret and shadow archive producing against its *metaphysical surface* and throughout the atomic universe that it projected, an ecstatic, avisual you. Cinema has extracted you from the universe and at the same time opened in you a universe. Every possible form of and formless you; a cinema about you, only you. Universal.

Notes

0. Universes

1. Jorge Luis Borges, "The Library of Babel," in *Labyrinths: Selected Stories and Other Writings*, ed. Donald A. Yates and James E. Irby, trans. James E. Irby (New York: New Directions, 1964), 57.

2. Walter Benjamin, "The Task of the Translator," in *Illuminations: Essays and Reflections*, ed. Hannah Arendt, trans. Harry Zohn (New York: Schocken, 1968), 70.

3. Borges, "Library of Babel," 53.

4. Ibid., 54.

5. Daniel Tiffany, *Toy Medium: Materialism and Modern Lyric* (Berkeley: University of California Press, 2000), 45.

6. Deleuze says: "The paradox of this pure becoming, with its capacity to elude the present, is the paradox of infinite identity (the infinite identity of both directions or senses at the same time—of future and past, of the day before and the day after, of more or less, of too much and not enough, of active and passive, and of cause and effect)" (Gilles Deleuze, *The Logic of Sense*, ed. Constantin V. Boundas, trans. Mark Lester with Charles Stivale [New York: Columbia University Press, 1990], 3).

7. Borges, "Library of Babel," 52.

8. Ibid., 51.

9. Ibid., 54–55.

10. Ibid., 55.

11. In his parable "Before the Law," Franz Kafka describes a structure of Law that exists for one individual only and no one else. "'Everyone strives to reach the

Law,' says the man, 'so how does it happen that for all these many years no one but myself has ever begged for admittance?' The doorkeeper recognizes that the man has reached his end, and, to let his failing senses catch the words, roars in his ear: 'No one else could ever be admitted here, since this gate was made only for you. I am now going to shut it'" (Franz Kafka, *The Complete Stories,* ed. Nahum N. Glatzer, trans. Willa Muir and Edwin Muir [New York: Schocken, 1971], 4).

12. Borges, "Library of Babel," 51.

13. Ibid.

14. Ibid., 52 (original emphasis).

15. Ibid., 58.

16. Nicolas Abraham and Maria Torok, "The Topography of Reality: Sketching a Metapsychology of Secrets," in *The Shell and the Kernel: Renewals of Psychoanalysis,* ed. and trans. Nicholas T. Rand (Chicago: University of Chicago Press, 1994), 157 (original emphasis). Abraham and Torok add in a footnote: "When considered as a metapsychological concept, the word 'reality' needs to be capitalized, especially since all other forms of reality presuppose and derive from it. The metapsychological Reality of the secret is a counterpart to the reality of the outside world; the negation of the one entails the refusal of the other" (157n).

17. Jacques Derrida, *Archive Fever: A Freudian Impression,* trans. Eric Prenowitz (Chicago: University of Chicago Press, 1996), 100.

18. Jacques Derrida, *The Gift of Death,* trans. David Wills (Chicago: University of Chicago Press, 1995), 92.

19. Derrida, *Archive Fever,* 3 (original emphases).

20. In *Archive Fever,* Derrida traces the etymology of the archive to its origins in the law, the house of the law, the place where the *archons* "recall the law and call on or impose the law" (2).

21. Ibid., 66. "We are *en mal d'archive:* in need of archives.... To be *en mal d'archive* can mean something else than to suffer from a sickness, from a trouble, or from what the noun *mal* might name. It is to burn with a passion" (91).

22. Jean-Claude Lebensztejn suggested this analogy. Carolyn Steedman discovers, in the figure of archive fever, an underlying historical and material pathogen. What Steedman calls "archive fever proper" was caused, she believes, by bacteriological agents that resided in the books themselves and were carried by dust into the bodies of scholars. She says: "Medical men like [John] Forbes and [Charles] Thackrah were able to provide physiological and psychological causes of the fevers of scholarship (lack of exercise, bad air, and its 'passions,' which were excitement and ambition). By the time that a bacteriological explanation for their fevers was available, 'the literary man' as a victim of occupational disease had disappeared as a category. Lacking bacteriological understanding of the dust that preoccupied them, early investigators did not consider the book, the very stuff of the scholar's life, as a potential cause of his fever. And yet the book and its components (leather binding, various glues and adhesives, paper and its edging, and decreasingly, parchments and vellums of various types) concentrated in one object many of the industrial hazards and diseases that were mapped out in the course of the century" (Carolyn Steedman, *Dust: The Archive and Cultural History* [New Brunswick, NJ: Rutgers University Press,

2001], 22). Gilles Deleuze says, "The task of perception entails pulverizing the world, but also one of spiritualizing its dust" (Gilles Deleuze, *The Fold: Leibniz and the Baroque,* trans. Tom Conley [Minneapolis: University of Minnesota Press, 1993], 87).

23. Derrida, *Gift of Death,* 21.

24. Derrida, *Archive Fever,* 64. "Comme si on ne pouvait pas, précisément, rappeler et archiver cela même qu'on refoule, l'archiver en le refoulant (car le refoulement est une archivation), c'est-à-dire archiver *autrement,* refouler l'archive en archivant le refoulement" (Jacques Derrida, *Mal d'archive: Une impression freudienne* [Paris: Galilée, 1995], 103, original emphasis).

25. Giorgio Agamben, *Remnants of Auschwitz: The Witness and the Archive,* trans. Daniel Heller-Roazen (New York: Zone, 1999), 144.

26. Derrida, *Gift of Death,* 60.

27. Agamben, *Remnants of Auschwitz,* 144.

1. The Shadow Archive (A Secret Light)

1. Jacques Derrida, *Archive Fever: A Freudian Impression,* trans. Eric Prenowitz (Chicago: University of Chicago Press, 1996), 34. "Freudian psychoanalysis proposes a new theory of the archive; it takes into account a topic and a death drive without which there would not in effect be any desire or possibility for the archive" (29).

2. Sigmund Freud, "Moses and Monotheism: Three Essays," in *The Standard Edition of the Complete Psychological Works of Sigmund Freud,* ed. and trans. James Strachey (London: Hogarth, 1955), 23:56.

3. Ibid. (emphasis added). Apparently, Freud had written the first essays four years earlier in 1934.

4. Jacques Derrida, *The Gift of Death,* trans. David Wills (Chicago: University of Chicago Press, 1995), 29–30.

5. Freud, "Moses and Monotheism," 7.

6. Ibid., 10–11.

7. Ibid., 16.

8. Ibid., 50.

9. Ibid., 50–51.

10. Ibid., 53.

11. Ibid., 54.

12. Ibid., 55.

13. Jacques Derrida, *Given Time: I. Counterfeit Money,* trans. Peggy Kamuf (Chicago: University of Chicago Press, 1992), 17.

14. Ibid., 14 (original emphasis).

15. For Derrida, "The archontic principle of the archive is also a principle of consignation, that is, of gathering together" (*Archive Fever,* 3). "*Consignation* aims to coordinate a single corpus, in a system or a synchrony in which all the elements articulate the unity of an ideal configuration" (3).

16. Derrida, *Archive Fever,* 11 (original emphasis).

17. Ibid., 19 (original emphasis).

18. Freud, "Moses and Monotheism," 57.

19. Ibid.

20. Ibid., 58 (original emphasis).

21. Ibid., 57.

22. Ibid., 58 (emphasis added). Agamben describes, in Giorgio Manganelli, the phenomenon of "homopseudonymy," which "consists in using a pseudonym that is in every respect identical to one's own name" (Giorgio Agamben, *Remnants of Auschwitz: The Witness and the Archive,* trans. Daniel Heller-Roazen [New York: Zone, 1999], 130). Like Freud's text, which is signed by him but written by another, estranged in the act of signing, the effect of a homopseudonymic mark is to tear the author from the text, but also from himself or herself. "The homopseudonym is absolutely foreign and perfectly intimate, both unconditionally real and necessarily non-existent, so much so that no language could describe it; no text could guarantee its consistency" (131). Derrida links the pseudonym to both secrecy and patronymy, claiming that all pseudonyms "are destined to keep secret the real name *as* patronym, the name of the father of the work, in fact the name of the father of the father of the work" (*Gift of Death,* 58).

23. Freud, "Moses and Monotheism," 66. In Freud's analysis, prescientific intuitions about solar power as the source of life serve as the first step toward monotheism. Of the Egyptian pharaoh, the young Amenophis IV, who embraced the religion of Aten and the worship of the sun, Freud says: "In an astonishing presentiment of later scientific discovery he recognized in the energy of solar radiation the source of all life on earth and worshipped it as a symbol of the power of his god" (59). Metaphors of light and power, darkness and secrecy, energy and radiation sustain the rhetorical structure of Freud's work.

24. Ibid., 70. Freud explains: "The phenomenon of latency in the history of the Jewish religion, with which we are dealing, may be explained, then, by the circumstance that the facts and ideas which were intentionally disavowed by what may be called the official historians were in fact never lost. Information about them persisted in traditions which survived among the people" (69). Traditions and the people who practice them serve as archives of the secret, secret archives where the disavowed idea lives in hiding, in secrecy, in latency.

25. Ibid., 70.

26. Ibid., 69.

27. Ibid., 70.

28. Ibid., 69.

29. Derrida, *Archive Fever,* 10 (original emphasis).

30. Ibid., 94. For Derrida, "Freudian psychoanalysis proposes a new theory of the archive; it takes into account a topic and death drive without which there would not in effect be any desire or any possibility for the archive" (29).

31. Ibid., 84.

32. Ibid., 86.

33. Freud, "Moses and Monotheism," 103.

34. Ibid., 105.

35. Derrida, *Archive Fever*, 2. "But it also *shelters* itself from this memory which it shelters," Derrida adds, "which comes down to saying also that it forgets it."

36. Jun'ichirô Tanizaki, *In Praise of Shadows*, trans. Thomas J. Harper and Edward G. Seidensticker (New Haven, CT: Leete's Island Books, 1977), 30.

37. Martin Heidegger, "The Question concerning Technology," in *The Question concerning Technology and Other Essays*, trans. William Lovitt (New York: Harper and Row, 1977), 20.

38. Tanizaki, *In Praise of Shadows*, 5.

39. Ibid.

40. Ibid., 11.

41. Ibid., 11–12.

42. Ibid., 36.

43. Ibid., 37.

44. Ibid.

45. Ibid.

46. Ibid., 35.

47. Ibid., 42.

48. Jacques Derrida, "NO APOCALYPSE, NOT NOW (full speed ahead, seven missiles, seven missives)," trans. Catherine Porter and Philip Lewis, *Diacritics* 14.2 (1984): 26.

49. Ibid., 28.

50. Ibid., 27. "If we are bound and determined to speak in terms of reference, nuclear war is the only possible referent of any discourse and any experience that would share their condition with that of literature" (28).

51. Ibid., 23.

52. Jacques Derrida, *Cinders*, trans. and ed. Ned Lukacher (Lincoln: University of Nebraska Press, 1991), 57.

53. Ibid., 73.

54. Jacques Derrida, "Passages—from Traumatism to Promise," in *Points . . . Interviews, 1974–1994*, ed. Elisabeth Weber, trans. Peggy Kamuf et al. (Stanford, CA: Stanford University Press, 1995), 391. "The difference between the trace 'cinder' and other traces is that the body of which cinders is the trace has totally disappeared, it has totally lost its contours, its form, its colors, its natural determination. Non-identifiable. And forgetting itself is forgotten. Everything is annihilated in the cinders" (391).

55. Derrida, *Cinders*, 37.

56. Derrida, *Archive Fever*, 1.

57. Ibid., 100.

58. Derrida, *Cinders*, 61.

59. Derrida, *Archive Fever*, 20 (original emphasis). Derrida is referring to an inscription, cited by Yosef Hayim Yerushalmi, in a Bible given to Freud by his father, Jakob, on the son's thirty-fifth birthday. Derrida reads the line "a cover of new skin" as a "sign of circumcision" (42). See Yosef Hayim Yerushalmi, *Freud's Moses: Judaism Terminable and Interminable* (New Haven, CT: Yale University Press, 1991).

60. Derrida, *Archive Fever*, 20.

61. Ibid., 42.

62. Ibid., 11.

63. In *The Gift of Death*, Derrida relates secrecy and death to paternity, specifically the relationship between a father and son, to a version of the gift, sacrifice. Referring to Abraham's gift to God of his own son Isaac, Derrida claims that this sacrifice, this gift belongs and returns both to Abraham and Isaac: "It is the sacrifice of both of them, it is the gift of death one makes to the other in putting *oneself* to death, mortifying oneself in order to make a gift of this death as a sacrificial offering to God" (69). This mode of sacrifice and secrecy, the gift of oneself and of death, marks for Derrida a certain impossibility of the gift of death between and of the father and son, on the very grounds that the paternal relationship is constituted. "Dying," he says, "can never be taken, borrowed, transferred, delivered, promised, or transmitted" (44).

64. Ibid., 53–55.

65. Derrida, *Archive Fever*, 26 (original emphasis).

66. Maurice Merleau-Ponty, "Eye and Mind," in *The Primacy of Perception: And Other Essays on Phenomenological Psychology, the Philosophy of Art, History, and Politics*, ed. James M. Edie, trans. Carleton Dallery (Evanston, IL: Northwestern University Press, 1964), 164 (emphasis added).

67. Derrida, *Gift of Death*, 88.

68. Ibid., 88–89.

69. "Quality, light, color, depth, which are there before us, are there only because they awaken an echo in our body and because the body welcomes them" (Merleau-Ponty, "Eye and Mind," 164).

70. Derrida, *Gift of Death*, 91.

71. Ibid., 89.

72. Ibid., 90.

73. Ibid.

74. Ibid.

75. Ibid.

76. Ibid., 89.

77. Ibid.

2. Modes of Avisuality

1. After a series of diagnoses and unsuccessful analyses, Freud had referred Emma Eckstein to his mentor-friend Wilhelm Fliess, a nasologist, who had performed an operation on Eckstein's nose, removing her turbinal bones. Eckstein had not recovered, and Freud began to have doubts, first about his own methods of analysis and then about the surgical skills of Fliess. His fear was confirmed when he discovered that Fliess had failed to remove "at least half a meter of gauze" from Eckstein's nose, which had become infected. Eckstein was hemorrhaging profusely and almost died before Freud discovered Fliess's malpractice. In a letter to Fliess dated 8 March 1895, Freud describes the horrendous discovery: "There was still

moderate bleeding from the nose and mouth; the fetid odor was very bad. Rosanes cleaned the area surrounding the opening, removed some sticky blood clots, and suddenly pulled at something like a thread, kept on pulling. Before either of us had time to think, at least half a meter of gauze had been removed from the cavity. The next moment came a flood of blood. The patient turned white, her eyes bulged, and she had no pulse. Immediately thereafter, however, he again packed the cavity with fresh iodoform gauze and the hemorrhage stopped. It lasted about half a minute, but this was enough to make the poor creature, whom we had lying flat, unrecognizable (Sigmund Freud, *The Complete Letters of Sigmund Freud to Wilhelm Fliess, 1887–1904*, ed. and trans. Jeffrey Moussaieff Masson [Cambridge, MA: Belknap Press of Harvard University Press, 1985], 116–17). The unrecognizable creature is for a moment dead: "She had no pulse." Freud discovers inside Eckstein's body a foreign object, discarded by Fliess. Freud insists to Fliess that his nausea at the scene of this episode comes not from the "flood of blood" but from the injustice: "So we had done her an injustice; she was not at all abnormal, rather a piece of iodoform gauze had gotten torn off as you were removing it and stayed in for fourteen days, preventing healing; at the end it tore off and provoked bleeding" (117). The affair is marked by an ethics of interiority and exteriority, of normalcy and abnormalcy, and the economy of a foreign body deposited in the cavity of Emma Eckstein's complex body.

2. Freud returned to the question of the border between psyche and body, internal and external stimuli repeatedly. In "Instincts and Their Vicissitudes" (1915), Freud offers the following psychocorporeal analogy: "Let us imagine ourselves in the situation of an almost entirely helpless living organism, as yet unorientated in the world, which is receiving stimuli in its nervous substance. This organism will very soon be in a position to make a first distinction and a first orientation. On the one hand, it will be aware of stimuli which can be avoided by muscular action (flight); these it ascribes to the external world. On the other hand, it will also be aware of stimuli against which such action is of no avail and whose character and constant pressure persists in spite of it; these stimuli are the signs of an internal world, the evidence of instinctual needs. The perceptual substance of the living organism will thus have found in the efficacy of its muscular activity a basis for distinguishing between an 'outside' and an 'inside'" (Sigmund Freud, "Instincts and Their Vicissitudes," in *The Standard Edition of the Complete Psychological Works of Sigmund Freud*, ed. and trans. James Strachey [London: Hogarth, 1957], 14:119). The genesis of an organism's recognition of inside and outside spaces follows the distinction between its corporeal (muscular) and psychic (instinctual) functions. Later in the same essay Freud says, "If we now apply ourselves to considering mental life from a *biological* point of view, an 'instinct' [*Trieb*] appears to us as *a concept on the frontier between the mental and the somatic,* as the psychical representative of the stimuli originating from within the organism and reaching the mind, as a measure of the demand made upon the mind for work in consequence of its connection with the body" (121–22, second emphasis added).

3. Sigmund Freud to Wilhelm Fliess, 12 June 1900, in Freud, *Complete Letters of Sigmund Freud to Wilhelm Fliess,* 417.

4. Sigmund Freud, *The Interpretation of Dreams,* in *The Standard Edition of the Complete Psychological Works of Sigmund Freud,* ed. and trans. James Strachey (London: Hogarth, 1958), 4:107.

5. Freud adds this commentary: "I might have said this to her in waking life.... It was my view at the time (though I have since recognized it was a wrong one) that my task was fulfilled when I informed patients of the hidden meaning of his symptoms" (*Interpretation of Dreams,* 108).

6. See Anne Friedberg, *Window Shopping: Cinema and the Postmodern* (Berkeley: University of California Press, 1993). Friedberg describes the reorganization of urban and psychic space that followed from the development of iron and glass architecture, which installed windows between private and public spaces. "The once-private interior became a public realm, the once-public exterior became privatized" (64). Seen in this light, Freud's dream window may reflect an architectonic desire, the exteriorization of his desire to see into the private interiority of Irma.

7. Like Tanizaki's exoskeletal house, Freud's carnal architecture, psychoarchitectural dream edifice, erases many of the borders between the body and the world, determining a form of visuality that is at once interior and exterior, private and public. Of the body and its relation to the world of visuality, Maurice Merleau-Ponty says, "Things have an internal equivalent in me; they arouse in me a carnal presence. Why shouldn't these [correspondences] in turn give rise to some [external] visible shape in which anyone else would recognize these motifs which support his own inspection of the world?" (Maurice Merleau-Ponty, "Eye and Mind," in *The Primacy of Perception: And Other Essays on Phenomenological Psychology, the Philosophy of Art, History, and Politics,* ed. James M. Edie, trans. Carleton Dallery [Evanston, IL: Northwestern University Press, 1964], 164).

8. Freud, *Interpretation of Dreams,* 107.

9. Ibid.

10. In his analysis of this section, Freud disavows his fantastic vision, noting only that it was common practice to examine adult women, as opposed to children, fully clothed. "Further than this," Freud says, "I could not see. Frankly, I had no desire to penetrate more deeply at this point" (ibid., 113).

11. Ibid., 107.

12. Ibid., 109.

13. Ibid.

14. In the dreamwork, all ideas can share equal value, since the force of any particular idea is given by the affect attached to it. Thus people, objects, words, sounds, and any other form of signifier can convey important aspects of the dream. Freud eventually locates a possible source of this dream displacement, suggesting that he might have wanted to substitute this other woman for Irma because "she [the other] would have *opened her mouth properly,* and have told me more than Irma" (ibid., 111, original emphasis). Wilhelm Fliess also haunts this dream and may be its secret subject.

15. Freud never found a satisfactory model for the activities of the psychic apparatus and actively criticized the demand for spatial or graphic figures of the

psyche. In *Civilization and Its Discontents,* Freud writes of the unconscious: "The fact remains that only in the mind is such a preservation of all the earlier stages alongside the final form possible, and that we are not in a position to represent this phenomenon in pictorial terms" (Sigmund Freud, "Civilization and Its Discontents," in *The Standard Edition of the Complete Psychological Works of Sigmund Freud,* ed. and trans. James Strachey [London: Hogarth, 1964], 21:71). The one exception may be the Mystic Writing-Pad or "*Wunderblock.*" See Sigmund Freud, "A Note Upon the 'Mystic Writing-Pad,'" in *The Standard Edition of the Complete Psychological Works of Sigmund Freud,* ed. and trans. James Strachey (London: Hogarth, 1961), 19:227–31.

16. Freud comments on this dream thought: "This, as may well be believed, is a perpetual source of anxiety to a specialist whose practice is almost limited to neurotic patients and who is in the habit of attributing to hysteria a great number of symptoms which other physicians treat as organic" (Freud, *Interpretation of Dreams,* 109). On this subject, Judith Butler says, "If there is a materiality of the body that escapes from the figures it conditions and by which it is corroded and haunted, then this body is neither a surface nor a substance, but the linguistic occasion of the body's separation from itself, one that eludes its capture by the figure it compels" (Judith Butler, "How Can I Deny That These Hands and This Body Are Mine?" in *Material Events: Paul de Man and the Afterlife of Theory,* ed. Tom Cohen, Barbara Cohen, J. Hillis Miller, and Andrzej Warminski [Minneapolis: University of Minnesota Press, 2001], 271–72).

17. Jacques Lacan, "The Dream of Irma's Injection," in *The Seminar of Jacques Lacan, Book II: The Ego in Freud's Theory and in the Technique of Psychoanalysis, 1954–1955,* ed. Jacques-Alain Miller, trans. Sylvana Tomaselli (New York: Norton, 1988), 154–55 (original emphasis).

18. In the idiom of Gilles Deleuze and Félix Guattari, Irma's formless, avisual, and inside-out body is organless, inorganic, a "body without an image" (Gilles Deleuze and Félix Guattari, *Anti-Oedipus: Capitalism and Schizophrenia,* trans. Robert Hurley, Mark Seem, and Helen R. Lane [Minneapolis: University of Minnesota Press, 1983], 8). "The body without organs is not the proof of an original nothingness, nor is it what remains of a lost totality. Above all it is not a projection; it has nothing whatsoever to do with the body itself, or with an image of the body" (8).

19. Trinh T. Minh-ha, "The World as a Foreign Land," in *When the Moon Waxes Red: Representation, Gender, and Cultural Politics* (New York: Routledge, 1991), 187.

20. Lacan, "Dream of Irma's Injection," 170.

21. Ernst Cassirer, *The Philosophy of the Enlightenment,* trans. Fritz C. A. Koelln and James P. Pettegrove (Princeton, NJ: Princeton University Press, 1951).

22. Max Horkheimer and Theodor Adorno, *Dialectic of Enlightenment,* trans. John Cumming (New York: Continuum, 1944), 6.

23. Gilles Deleuze, *The Logic of Sense,* ed. Constantin V. Boundas, trans. Mark Lester with Charles Stivale (New York: Columbia University Press, 1990), 87.

24. Ibid., 87 (emphasis added).

25. Catherine Waldby, *The Visible Human Project: Informatic Bodies and Posthuman Medicine* (London: Routledge, 2000), 91.

26. Linda Dalrymple Henderson, "X Rays and the Quest for Invisible Reality in the Art of Kupka, Duchamp, and the Cubists," *Art Journal* 47.4 (1988): 324. See also Linda Dalrymple Henderson, "A Note on Francis Picabia, Radiometers, and X Rays in 1913," *Art Bulletin* 71.1 (1989): 114–23; W. Robert Nitske, *The Life of Wilhelm Conrad Röntgen, Discoverer of the X Ray* (Tucson: University of Arizona Press, 1971).

27. Horkheimer and Adorno, *Dialectic of Enlightenment,* 3 (emphasis added).

28. Richard F. Mould, *A Century of X-Rays and Radioactivity in Medicine: With Emphasis on Photographic Records of the Early Years* (Bristol, PA: Institute of Physics Publishing, 1993), 1.

29. Lisa Cartwright, "Women, X-rays, and the Public Culture of Prophylactic Imaging," *Camera Obscura* 29 (May 1992): 30. A revised version of Cartwright's essay appears in her study of medical imagery, *Screening the Body: Tracing Medicine's Visual Culture* (Minneapolis: University of Minnesota Press, 1995). Operating simultaneously in the fields of film history, medical rhetoric and culture, and feminist theory and criticism, Cartwright's searching analysis situates the X-ray at the crossroads of twentieth-century arts and sciences. She writes: "The X ray is the pivotal site of investigation for this book's exploration of medicine's technological/visual knowledge, desire, and power. It is also the most conflicted site, embodying multiple paradigms of visuality and multiple political agendas" (*Screening the Body,* 108).

30. Cartwright, *Screening the Body,* 115. Hands, says Mould, were also among the most vulnerable areas of the human body, where many of the earliest signs of X-ray injuries to pioneer physicians and technicians first appeared. Röntgen's own hands were unscathed. Alongside a photograph of a plaster cast made of Röntgen's hands "immediately after his death in 1923," Mould notes that "unlike the hands of other X-ray pioneers, it is seen that Röntgen's are undamaged" (*Century of X-Rays,* 5).

31. Cartwright, *Screening the Body,* 115.

32. Ibid.

33. Otto Glasser, *Wilhelm Conrad Röntgen and the Early History of the Roentgen Rays* (Springfield, IL: Thomas, 1934), 81. "Many people," according to Glasser, "reacted strongly to the ghost pictures. The editor of the *Grazer Tageblatt* had a roentgen picture taken of his head and upon seeing the picture 'absolutely refused to show it to anybody but a scientist. He had not closed an eye since he saw his own death's head'" (81).

34. Lacan, "Dream of Irma's Injection," 163–64.

35. Henderson, "X Rays and the Quest for Invisible Reality," 324.

36. Ibid., 336. A similar pattern of immersion in the fantastic properties of unknown fluids followed Marie Curie's discovery of radium in 1898, in the wake of Henri Becquerel's discovery of radioactivity in 1896. The same imaginary properties that were attributed to X-rays accrued to radium, which was seen as an elixir of life, as the source of life itself. A frantic effort to introduce radium into the body ensued, flooding the marketplace with radioactive commodities and services: toothpaste, cocktails, spas, as well as bug sprays, which also cleaned and polished furniture. In pure form, bottled *Agua Radium* allowed the fastest means of ingestion. After the initial euphoria began to fade, the toxic but also photographic effects of radioactivity began to appear, glowing. At watch factories where women painted the hands and

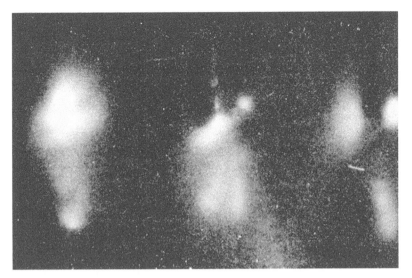

Radioactive mouth.

dials of watches with radium, a practice known as "tipping" (licking the brushes to form a point at the tip) served as a method for transporting the radium into their bodies. Like a Cheshire cat, the image of a woman's mouth survives. Her teeth absorbed so much radium she could develop film in her mouth. Her radioactive mouth had become a camera lab, a *camera dentata*. Pierre Curie himself, says Mould, hoped that "radium might help restore eyesight to the blind" (*Century of X-Rays*, 21). See Claudia Clark, *Radium Girls: Women and Industrial Health Reform, 1910–1935* (Chapel Hill: University of North Carolina Press, 1997).

37. Waldby, *Visible Human Project,* 14. "The Visible Woman," Waldby notes, "represents a technical improvement on the Man: the body was planed into much finer cross-sections (5,189 sections [as opposed to 1,878 sections of the Man]) which produced higher resolution in the resulting images, and a much larger data file" (15).

38. Lisa Cartwright, "A Cultural Anatomy of the Visible Human Project," in *The Visible Woman: Imagining Technologies, Gender, and Science,* ed. Paula A. Treichler, Lisa Cartwright, and Constance Penley (New York: New York University Press, 1998), 39.

39. Waldby, *Visible Human Project,* 13.

40. Ibid., 63.

41. Ibid., 63 (original emphasis).

42. Ibid., 64.

43. Ibid., 5.

44. Ibid., 92 (original emphasis).

45. Ibid., 159.

46. Ibid., 14.

47. William Haver, *The Body of This Death: Historicity and Sociality in the Time of AIDS* (Stanford, CA: Stanford University Press, 1996), 55. Haver is following the

idiom and tone of Derrida's "NO APOCALYPSE, NOT NOW (full speed ahead, seven missiles, seven missives)," trans. Catherine Porter and Philip Lewis, *Diacritics* 14.2 (1984): 26.

48. Haver, *Body of This Death*, 55 (original emphasis). The apocalypse is framed, in Haver's discussion, by the atomic bombings of Hiroshima and Nagasaki and the AIDS pandemic, by "the force of an Outside that is not merely the outside of an inside, but the outside that is inside, the insidious inside" (53).

49. Roland Barthes, *Camera Lucida: Reflections on Photography*, trans. Richard Howard (New York: Hill and Wang, 1981). Barthes writes: "A photograph's *punctum* is *that accident* which pricks me (but also bruises me, is poignant to me)" (26–27, emphasis added).

50. Victor Bouillion, "War and Medicinema: The X-ray and Irradiation in Various Theaters of Operation," in *Incorporations*, ed. Jonathan Crary and Sanford Kwinter (New York: Zone, 1992), 253.

51. Concerning the partnership between the field of physics and X-ray technology during the twentieth century, the decision of the 1994 Nobel Prize committee may have signaled a critical moment. Among the recipients of the awards in science were two physicists, Clifford G. Shull and Bertram N. Brockhouse, who had succeeded in developing "neutron probes [that] gave scientists a set of tools more powerful than X-rays and other forms of radiation used for exploring the atomic structure of matter" (Malcolm W. Brown, "American Awarded Nobel Prize in Chemistry," *New York Times*, 13 October 1994). X-rays may have been superseded, even rendered obsolete, on the eve of the centennial of their discovery. Röntgen himself was the recipient of the first Nobel Prize for science in 1901.

52. Félix Nadar, *Nadar: Dessins et Écrits*, 2 vols. (Paris: Hubschmid, 1979), 2:978 (my translation). "Donc selon Balzac, chaque corps dans la nature se trouve composé de séries de spectres, en couches superposées à l'infini, foliacées en pellicules infinitésimales, dans tous les sens où l'optique perçoit ce corps. L'homme à jamais ne pouvant créer,—c'est à dire d'une apparition, de l'impalpable, constituer une chose solide, ou de *rien* faire une *chose*,—chaque opération Daguerrienne venait donc surprendre, détachait et retenait en se l'appliquant une des couches du corps objecté."

53. A version of this phobia also existed in Meiji Japan (1868–1912). "During this time," writes Sakuma Rika, "numerous superstitions were associated with photography, including the beliefs that 'posing for a photograph drained one's shadow,' 'posing for a second shortened one's life,' and 'when three subjects posed for a photograph, the one in the middle would die'" (Sakuma Rika, "Shashin to josei—atarashî shikaku media no tôjô to 'miru/mirareru' jibun no shutsugen" ["Photography and Women: The Advent of New Visual Media and the Creation of a Self That Looks/Is Looked At"], in *Onna to otoko no jikû: semegiau onna to otoko* [*The Timespace of Women and Men: The Confrontation of Women and Men*], vol. 5 of *Nihon joseishi saikô* [*Redefining Japanese Women's History*], ed. Okuda Akiko [Tokyo: Fujihara Shoten, 1995], 200 (my translation). Regarding the photoiconography of hands and the image of Berthe Röntgen's left hand, many Japanese women at that time feared having their hands photographed. Small hands were considered a sign of feminine beauty, and, according to Sakuma, these women believed that their hands not only

looked bigger in photographs but that the photographic process actually made their hands swell (228).

54. See Akira Mizuta Lippit, "Photographing Nagasaki: From Fact to *Artefact*," in *Nagasaki Journey: The Photographs of Yosuke Yamahata, August 10, 1945,* ed. Rupert Jenkins (San Francisco: Pomegranate, 1995), 25–29.

55. Michel Frizot, "The All-Powerful Eye: The Forms of the Invisible," in *A New History of Photography*, ed. Michel Frizot, trans. Susan Bennett, Liz Clegg, John Crook, and Caroline Higgitt (Cologne: Könemann, 1998), 281.

56. Glasser, *Wilhelm Conrad Röntgen,* 82.

57. Ibid., 83.

58. Daniel Tiffany, *Radio Corpse: Imagism and the Cryptaesthetic of Ezra Pound* (Cambridge, MA: Harvard University Press, 1995), 226. "On the one hand," says Tiffany of the relation between radium and the X-ray, "we have a substance whose radiant energy is invisible, whereas on the other we have a form of radiant energy that produces images of the unseen" (226).

59. László Moholy-Nagy, *Vision in Motion* (Chicago: Theobald, 1947), 252.

60. Cartwright, *Screening the Body,* 113.

61. David Sylvester, *The Brutality of Fact: Interviews with Francis Bacon* (Oxford: Thames and Hudson, 1987). Of the photograph, Bacon, who claims to have been influenced by radiographic images, says: "I think it's the slight remove from fact, which returns me onto the fact more violently" (30).

62. In *Camera Lucida,* Barthes says: "A specific photograph, in effect, is never distinguished from its referent.... It is as if the Photograph always carries its referent with itself, both affected by the same amorous or funereal immobility, at the very heart of the moving world.... The Photograph belongs to that class of laminated objects whose leaves cannot be separated without destroying them both" (5–6).

63. Moholy-Nagy, *Vision in Motion,* 252.

64. Tiffany cites "Pound's fascination with radium and radioactivity" as an example of the "negativity of the modernist Image—a negativity that is signaled by the Image's resistance to visuality" (*Radio Corpse,* 226).

65. Jacques Derrida, "Différance," in *Margins of Philosophy,* trans. Alan Bass (Chicago: University of Chicago Press, 1982), 24 (emphasis added).

66. François Dagognet, *Etienne-Jules Marey: A Passion for the Trace,* trans. Robert Galeta with Jeanine Herman (New York: Zone, 1992).

67. Richard Crangle says that in the British press, for example, excited discussions in 1896 about the "New Photography" referred not to the Cinématographe but to the X-ray. "The interest surrounding the New Photography in early 1896 seems, a century later, to have eclipsed the almost parallel launch in Britain of an optical sensation which became far more influential: the projected moving picture" (Richard Crangle, "Saturday Night at the X-rays—The Moving Picture and 'The New Photography' in Britain, 1896," in *Celebrating 1895: The Centenary of Cinema,* ed. John Fullerton [Sydney: Libbey, 1998], 138).

68. Paula Dragosh suggested the relationship between the logic of reversibility and anniversaries, "a chiasmus, after the Greek name for the letter 'x'" (letter to author, June 2005).

69. Wilhelm Conrad Röntgen, "On a New Kind of Rays (Preliminary Communication)," in Glasser, *Wilhelm Conrad Röntgen*, 41–52. The article was reprinted in the *Annual Report of the Smithsonian Institution* in 1897.

70. Other advances in film technology during 1895 include the introduction of R. W. Paul and Brit Acres's movie camera in March, the advent of the so-called Latham loop by Woodville Latham and his sons in April, and the debut of C. Francis Jenkins and Thomas Armat's Phantoscope in October. Thomas Alva Edison's "Kinetoscope" and "Vitascope" exhibitions straddled 1895, taking place in 1894 and 1896, respectively.

71. Emmanuelle Toulet, *Birth of the Motion Picture*, trans. Susan Emanuel (New York: Abrams, 1995), 40.

72. According to Bertrand Tavernier's narration on the Kino Video collection of the Lumières' films, *The Lumière Brothers' First Films* (1996).

73. Noël Burch, *Life to Those Shadows*, ed. and trans. Ben Brewster (Berkeley: University of California Press, 1990), 15.

74. Josef Breuer and Sigmund Freud, *Studies on Hysteria,* in *The Standard Edition of the Complete Psychological Works of Sigmund Freud,* ed. and trans. James Strachey (London: Hogarth, 1955), vol. 2.

75. Sigmund Freud, "The Metapsychological Supplement to the Theory of Dreams," in *The Standard Edition of the Complete Psychological Works of Sigmund Freud,* ed. and trans. James Strachey (London: Hogarth, 1957), 14:223 (original emphasis). "A dream tells us," Freud writes, "that something was going on which tended to interrupt sleep, and it enables us to understand in what way it has been possible to fend off this interruption. The final outcome is that the sleeper has dreamt and is able to go on sleeping; the internal demand which was striving to occupy him has been replaced by an external experience, whose demand has been disposed of. A dream is therefore, among other things a *projection:* an externalization of an internal process."

76. Henderson, "X rays and the Quest for Invisible Reality," 325.

77. Lisa Cartwright and Brian Goldfarb, "Radiography, Cinematography and the Decline of the Lens," in *Incorporations* (New York: Zone, 1992), 190–201.

78. Moholy-Nagy, *Vision in Motion,* 210. See in this connection Thomas Mann, *The Magic Mountain,* trans. H. T. Lowe-Porter (New York: Knopf, 1951).

79. Walter Benjamin, "The Work of Art in the Age of Mechanical Reproduction," in *Illuminations,* ed. Hannah Arendt, trans. Harry Zohn (New York: Schocken, 1968), 236–37 (emphasis added).

80. Addendum to 1895: Albert de Rochas published in 1895 his collection of spectral and psychic images, *L'Exteriorisation de la sensibilité: Étude experimentale et historique in Paris.* Also in 1895, Hippolyte Baraduc began work on psychic photography—accomplished by inducing the subject to excrete "psycho-odo-fluidiques" that were then captured on the photographic plate. His *L'Âme humaine, ses mouvements, ses lumières et l'iconographie de l'invisible* appeared the following year (Paris: Carré, 1896). Baraduc, a gynecologist, believed that the human soul emitted a "subtle force" only perceptible to the camera. The specter of women's interiority is again invoked to provide a figure for the representation of the invisible. Regarding

the popularity of spirit photography that erupted in the late nineteenth century, Tom Gunning writes: "The medium herself became a sort of camera, her spiritual negativity bodying forth a positive image, as the human body behaves like an uncanny photomat, dispensing images from its orifices" (Tom Gunning, "Phantom Images and Modern Manifestations: Spirit Photography, Magic Theater, Trick Films, and Photography's Uncanny," in *Fugitive Images: From Photography to Video,* ed. Patrice Petro [Bloomington: Indiana University Press, 1995], 58).

81. Jacques Derrida, *Of Grammatology,* trans. Gayatri Chakravorty Spivak (Baltimore, MD: Johns Hopkins University Press, 1976), 70.

3. Cinema Surface Design

1. Germaine Dulac, "Visual and Anti-visual Films," in *The Avant-Garde Film: A Reader of Theory and Criticism,* ed. P. Adams Sitney, trans. Robert Lamberton (New York: Anthology Film Archives, 1987), 31.

2. Ibid.

3. Ibid., 32.

4. Maurice Merleau-Ponty, "Eye and Mind," in *The Primacy of Perception: And Other Essays on Phenomenological Psychology, the Philosophy of Art, History, and Politics,* ed. James M. Edie, trans. Carleton Dallery (Evanston, IL: Northwestern University Press, 1964), 170.

5. Laura U. Marks, *Touch: Sensuous Theory and Multisensory Media* (Minneapolis: University of Minnesota Press, 2002), 8. See also Laura U. Marks, *The Skin of the Film: Intercultural Cinema, Embodiment, and the Senses* (Durham, NC: Duke University Press, 2000).

6. Dulac, "Visual and Anti-visual Films," 32.

7. Germaine Dulac, "The Essence of Cinema: The Visual Idea," in Sitney, *Avant-Garde Film,* 37.

8. Dulac, "Visual and Anti-visual Films," 32.

9. Walter Benjamin, "The Work of Art in the Age of Mechanical Reproduction," in *Illuminations,* ed. Hannah Arendt, trans. Harry Zohn (New York: Schocken, 1968), 236.

10. W. K. L. Dickson and Antonia Dickson, *History of the Kinetograph, Kinetoscope, and Kinetophonograph* (New York: Museum of Modern Art, 2000), 43. W. K. L. Dickson was himself a key figure in the development of early cinema and worked with Thomas Alva Edison on the kinetograph, kinetoscope, and kinetophonograph.

11. Ibid.

12. Ibid.

13. Dulac, "Visual and Anti-visual Films," 32.

14. Tom Gunning, "An Aesthetic of Astonishment: Early Film and the (In) Credulous Spectator," in *Film Theory and Criticism: Introductory Readings,* ed. Leo Braudy and Marshall Cohen (New York: Oxford University Press, 1999), 819.

15. Ibid., 822. The Lumière episode is cited in Thomas Edison's 1902 film, *Uncle Josh at the Moving Picture Show* (directed by Edwin S. Porter), which features a film within a film. At the moving picture show, Uncle Josh reacts to a series of

films as if they were real, recoiling in fright to the film of an oncoming train. Edison's film ends when the screen is torn down and the projectionist is exposed behind it.

16. Gunning refers to the myth as a "primal scene," which continues to haunt later theories of spectatorship: "The terrorized spectator of the Grand Café still stalks the imagination of film theorists who envision audiences submitting passively to an all-dominating apparatus, hypnotized and transfixed by its illusionist power" ("Aesthetic of Astonishment," 819).

17. Ibid., 820.

18. From the *New York Mail and Express,* 25 September 1897, reprinted in Kemp R. Niver, *The Biograph Bulletins, 1896–1908* (Los Angeles: Locare Research Group, 1971), 27. Cited in Gunning, "Aesthetic of Astonishment," 829. See Tom Gunning, "An Unseen Energy Swallows Space: The Space in Early Film and Its Relation to American Avant-Garde Film," in *Film before Griffith,* ed. John Fell (Berkeley: University of California Press, 1983), 355–66. Here, "an unseen energy swallows space."

19. For Deleuze, forces make representation possible; they drive representation, but are themselves outside representation, nonrepresentational, unrepresentable. See "Painting Forces" in *Francis Bacon: The Logic of Sensation,* trans. Daniel W. Smith (Minneapolis: University of Minnesota Press, 2003), 48–54.

20. Gunning, "Aesthetic of Astonishment," 827 (emphasis added).

21. Ibid., 826.

22. Michael Fried, *Absorption and Theatricality: Painting and Beholder in the Age of Diderot* (Chicago: University of Chicago Press, 1980), 50. Fried is referring to the effect of duration in absorption, the "illusion of imminent or gradual or even fairly abrupt change" (50). The still image is about to move, its stillness an effect of "time *filled,*" says Fried (51, original emphasis). In this sense, the transition from still image to moving image, which the Lumières performed at their early screenings, evokes the experience of time Fried describes as absorption.

23. A number of the early films discussed here are available on the Kino Video collection, *The Movies Begin,* 4 vols. (New York: Kino International Corporation, 2002). Program notes are by Charles Musser.

24. Reversibility seems to have been a principle of cinema from the outset, recognized by the Lumières, who projected *Demolition of a Wall* (*Démolition d'un mur,* no. 40, 1895) first forward, then backward, first destroying then restoring the wall.

25. The "deflation of space," says Sobchack of this phenomenon in science fiction films, "is presented not as a loss of dimension, but rather as an excess of surface" (Vivian Sobchack, *Screening Space: The American Science Fiction Film* [New York: Ungar, 1987], 256).

26. An inverse projectile pierces Georges Méliès's 1902 film *A Trip to the Moon (Le voyage dans la lune),* when a rocket penetrates the moon's surface, literally the personified face of the moon in its eye. The face of the moon, and in particular the locus of its eye, can be seen as a figure of the spectator, a displaced site of encounter between the spectator and cinema, like the sliced eye in Luis Buñuel and Salvador Dalí's 1928 film *Un Chien Andalou.*

27. Several other early films collected in *The Movies Begin* suggest folded spaces, other spaces within or behind visible spaces, and spaces *mise-en-abîme*. In James Williamson's 1901 *Stop Thief!* a thief is chased into a large barrel, where he seeks to hide from his pursuers. Four dogs follow the thief into the barrel, where they appear to attack him out of view. Eventually the robbed man arrives and removes the dogs one by one from the barrel, using cuts to create the effect of a dog pack in the barrel with the thief. The on-screen but folded interiority of the barrel becomes in *Stop Thief!* an expanded and virtual space into which the thief and four dogs are deposited. On-screen but unseen, the barrel represents an avisual interior filmic space.

Cecil Hepworth's 1905 film *The Other Side of the Hedge* (directed by Lewis Fitzhamon) deploys a hedge as a second screen or plane within the diegetic space, not unlike the factory wall of the Lumières' *Leaving the Lumière Factory*, which allows an eager couple to evade the surveillance of a chaperone. Their amorous pursuit is revealed only to the spectator through a 180-degree cut, which shifts the point of view to the other side, where they have staged a ruse—hats propped on sticks, which are visible from the side of the chaperone and appear to be set apart at a safe distance—to further delay discovery. The hedge functions perhaps as a metonymy of the screen, the other side as a fantastic other space, replete with restricted pleasures.

28. Gilles Deleuze, *The Logic of Sense,* ed. Constantin V. Boundas, trans. Mark Lester with Charles Stivale (New York: Columbia University Press, 1990), 125 (original emphasis). Deleuze also describes the "metaphysical surface" as a "surface of pure thought" (208) and as a "second screen" (221).

29. Deleuze's "metaphysical surface," which establishes spatial continuities where there are none—"a physics which endlessly assembles the variations and pulsations of the entire universe"—is a fundamental feature of what Charles Musser calls "screen practice," an expanded field of cinema that predates the invention of the basic apparatus. Of the magic lantern shows popular prior to the advent of cinema, Musser says, "Spatial continuities became important in the later part of the nineteenth century. Surviving documentation, some as early as 1860, indicates that in sequencing photographic views, practitioners were often preoccupied with the creation of a spatial world" (Charles Musser, *The Emergence of Cinema: The American Screen to 1907* [Berkeley: University of California Press, 1990], 38). A spatial world emerges from screen practice, one that involves at this moment, according to Musser, "frequent dissolves from exterior to interior" (38). This world to that, here to there, this side to that through a metaphysical screen practice.

30. *Interior New York Subway* is collected on *Treasures from American Film Archives: Fifty Preserved Films,* 4 vols. (National Film Preservation Foundation, 2000). Program notes are by Scott Simmon, notes on the music by Martin Marks. Bitzer is best known for his later collaborations with D. W. Griffith, especially for his action cinematography in Griffith's 1915 *Birth of a Nation.*

31. In his "Program Notes" for *The Movies Begin,* Musser points out the popularity of films depicting trains plunging into tunnels, "phantom rides," following Biograph's 1897 *Haverstraw Tunnel.*

32. In the "Program Notes" for *Treasures from American Film Archives,* Simmon adds: "The visual syncopation of the pillars in the subway tunnel was difficult to record—both on film and on our transfer to video—and was a cause for concern when the subway opened. 'Is subway travel injurious to the eyes?' the *New York Times* asked in its opening day reportage, and it provided a practical answer: 'A well known oculist says that looking at the rows of white columns is very straining. Therefore don't look at them'" (97). Simmon's account aligns the visuality of cinema with that of trains, especially with regard to rapid, repetitive, high contrast, and abstract imagery. *Interior New York Subway* also prefigures the work of experimental filmmakers like Peter Kubelka, Tony Conrad, and Paul Sharits, among others, who experimented with flicker effects. Ernie Gehr's 1970 *Serene Velocity* in particular echoes the illusory depths and movements of *Interior New York Subway.*

33. Alfred Hitchcock uses this technique of editing against dark screens, cuts on dark frames, to produce the illusion of a continuity—one uninterrupted shot—in *Rope* (1948).

34. Siegfried Kracauer says: "Any huge close-up reveals new and unsuspected formations of matter; skin textures are reminiscent of aerial photographs, eyes turn into lakes or volcanic craters. Such images blow up our environment in a double sense: they enlarge it literally; and in doing so, they blast the prison of conventional reality, opening up expanses which we have explored at best in dreams before" (Siegfried Kracauer, *Theory of Film: The Redemption of Physical Reality* [New York: Oxford University Press, 1965], 48). In Kracauer's rhetoric, extreme proximity erases the distinction between interiority and exteriority; X-rays, dreams, and the world come to share a common geography.

35. André Bazin, "Painting and Cinema," in *What Is Cinema?* ed. and trans. Hugh Gray (Berkeley: University of California Press, 1967), 1:166 (emphasis added).

36. Deleuze, *Logic of Sense,* 223.

37. Nicolas Abraham and Maria Torok locate the mouth as a critical locus for the processing of loss. Its dual function as the place of eating and speaking renders it the passage through which the literal (food) and figurative (language) are confused in the face of loss. At the moment of insurmountable loss, they say, the mouth reverts to the site of "antimetaphor." The mouth loses its capacity to speak, turning words and everything else in the world into some form of food. This moment marks the transition from mourning to melancholia, introjection to incorporation: "The crucial move away from introjection (clearly rendered impossible) to incorporation is made when *words* fail to fill the subject's void and hence an imaginary thing is inserted in their place. The desperate ploy of filling the mouth with illusory nourishment has the equally illusory effect of eradicating the idea of a void to be filled with words" (Nicolas Abraham and Maria Torok, "Mourning *or* Melancholia: Introjection *versus* Incorporation," in *The Shell and the Kernel: Renewals of Psychoanalysis,* ed. and trans. Nicholas T. Rand [Chicago: University of Chicago Press, 1994], 128–29, original emphasis). Seen in the light of Abraham and Torok's theory of radical loss, *The Big Swallow,* a silent film about losing oneself inside an expanding mouth, suggests a comedy based on an anxiety

of early cinema: demetaphorization, the transformation of representation into reality, and the danger of one's disappearance in it.

38. Deleuze, *Logic of Sense,* 9. Deleuze is following here the logic and trajectory of Lewis Carroll's *Alice's Adventures in Wonderland* and *Through the Looking-Glass.* Deleuze describes Alice's adventure as singular: a "climb to the surface" and "disavowal of false depth," "her discovery that everything happens at the border" (9).

39. Merleau-Ponty, "Eye and Mind," 180 (original emphasis).

40. Noël Burch, "A Primitive Mode of Representation?" in *Early Cinema: Space, Frame, Narrative,* ed. Thomas Elsaesser with Adam Barker (London: British Film Institute, 1990), 221. Burch insists on a spatial formulation in establishing a system specific to the primitive mode, one that is formed through what he calls an irreducible "primitive externality" (220).

41. Deleuze says of the phantasm, its mobility and geography: "It belongs as such to an ideational surface over which it is produced. It transcends inside and outside, since its topological property is to bring 'its' internal and external sides into contact, in order for them to unfold onto a single side" (Deleuze, *Logic of Sense,* 211).

42. Ibid., 203.

43. Sigmund Freud, "Beyond the Pleasure Principle," in *The Standard Edition of the Complete Psychological Works of Sigmund Freud,* ed. and trans. James Strachey (London: Hogarth, 1955), 18:24.

44. Ibid.

45. Ibid.

46. Ibid., 26.

47. Ibid., 27.

48. Ibid.

49. Ibid.

50. Albert Liu, letter to author, March 2004.

51. Deleuze, *Logic of Sense,* 213.

52. Ibid., 199.

53. Ibid., 217.

54. "Cinema, in contrast to sound recording," says Friedrich Kittler, "began with reels, cuts, and splices" (Friedrich A. Kittler, *Gramophone, Film, Typewriter,* trans. Geoffrey Winthrop-Young and Michael Wutz [Stanford, CA: Stanford University Press, 1999], 115).

55. Deleuze, *Logic of Sense,* 155.

56. Ibid.

57. Ibid., 156.

58. Sigmund Freud, "The Ego and the Id," in *The Standard Edition of the Complete Psychological Works of Sigmund Freud,* ed. and trans. James Strachey (London: Hogarth, 1961), 19:25 (original emphases). Freud continues by describing pain as a sensation through which new knowledge about the inside of one's body arrives. "Pain, too, seems to play a part in the process, and the way we gain new knowledge of our organs during painful illness is perhaps a model of the way by which in general we arrive at the idea of our body" (25–26).

59. Ibid., 26.

60. Ibid., 26n. Freud mentions this point of contact between the *Pcpt.-Cs.* and the brain in "Moses and Monotheism: Three Essays" (1939). While describing the topographical configuration of the psychical apparatus, Freud appears to disavow the relation between psyche and anatomy: "I will add the further comment that the psychical topography I have developed here has nothing to do with the anatomy of the brain, *and actually only touches it at one point*" (Sigmund Freud, "Moses and Monotheism: Three Essays," in *The Standard Edition of the Complete Psychological Works of Sigmund Freud,* ed. and trans. James Strachey [London: Hogarth, 1964], 23:97; emphasis added). The two systems do touch at only one point.

61. Freud, "The Unconscious," in *The Standard Edition of the Complete Psychological Works of Sigmund Freud,* ed. and trans. James Strachey (London: Hogarth, 1957), 14:187.

62. Bertram Lewin, "Sleep, the Mouth, and the Dream Screen," *Psychoanalytic Quarterly* 15 (1946): 419–43.

63. Jean-Louis Baudry, "The Apparatus: Metapsychological Approaches to the Impression of Reality in Cinema," in *Narrative, Apparatus, Ideology: A Film Theory Reader,* ed. Philip Rosen, trans. Jean Andrews and Bertrand Augst (New York: Columbia University Press, 1986), 310.

64. Dulac, "Visual and Anti-visual Films," 34.

65. Antonin Artaud, "Cinema and Reality," in *Antonin Artaud: Selected Writings,* ed. Susan Sontag, trans. Helen Weaver (New York: Farrar, Straus and Giroux, 1976), 151.

66. Deleuze, *Logic of Sense,* 103–4 (original emphasis).

67. Ibid., 93.

4. An Atomic Trace

1. Willem de Kooning, "What Abstract Art Means to Me" (1951), in *Collected Writings,* ed. George Scrivani (New York: Hanuman, 1988), 60. Of perverts and their angelic fantasies, Gilles Deleuze asks: "Why does the pervert have the tendency to imagine himself as a radiant angel, an angel of helium and fire?" (Gilles Deleuze, *The Logic of Sense,* ed. Constantin V. Boundas, trans. Mark Lester with Charles Stivale [New York: Columbia University Press, 1990], 319).

2. The unnamed protagonist of Chris Marker's 1962 film *La Jetée* survives a nuclear war by clinging to the memory of a woman he had seen just before the bombs struck. Aided by this image, which the film's narrator describes as "a scar," the protagonist travels back through time to be with her. Marker's film is composed for the most part of still images, which make the transition from present to past, memory to reality, and image to life, a movement from one static image to another.

3. Hollywood produced several versions of the invisible man (and woman) films before and during the war, from 1933 to 1944. The most famous of these is James Whale's 1933 adaptation of H. G. Wells's novel *The Invisible Man,* which features Claude Rains. Others include *The Invisible Man Returns* (1940), *The Invisible Woman* (1941), *Invisible Agent* (1942), and *The Invisible Man's Revenge* (1944).

4. For more on imposed and internal censorship in postwar Japanese cinema, see Kyoko Hirano's *Mr. Smith Goes to Tokyo: Japanese Cinema under the American Occupation, 1945–1952* (Washington, DC: Smithsonian Institution Press, 1992), especially chapter 2, "Prohibited Subjects," 47–103.

5. In the registers of postatomic urban planning and architecture, Edward Dimendberg describes the imperative to disperse as a key element of urban defense against nuclear assault. "Postwar urban planners noted that of the war's two atomic weapon targets, the number of deaths at Nagasaki were half those at Hiroshima, a consequence of the dispersed urban population. 'In an atomic war, congested cities would become death traps,' write Edward Teller and two fellow nuclear physicists in *The Bulletin of Atomic Scientists* in 1946. . . . Fear of density pervaded much planning discourse of the early 1950s and in its most extreme form equated urban concentration *tout court* with susceptibility to military attack" (Edward Dimendberg, "City of Fear: Defensive Dispersal and the End of Film Noir," *Any* 18 [1997]: 15). The dialectic of concentration and dispersal, excess visibility and vulnerability, seems to circulate from body to city in the immediate aftermath of the atomic war.

6. The relationship between extreme eros and the pursuit of science figures prominently, according to Jerome F. Shapiro, in atomic bomb cinema. He says, "In such films there is usually a scientist who postpones marriage to devote himself to scientific experiments that usurp the laws of nature" (Jerome F. Shapiro, *Atomic Bomb Cinema* [New York: Routledge, 2002], 45).

7. Mitsuhiro Yoshimoto, *Kurosawa: Film Studies and Japanese Cinema* (Durham, NC: Duke University Press, 2000), 195.

8. Teshigahara's film, based on Abe Kôbo's 1964 novel about a faceless man, also references Georges Franju's 1960 *Eyes without a Face (Les yeux sans visage).*

9. H. G. Wells, *The Invisible Man,* in *The Time Machine and the Invisible Man* (New York: Signet, 1984), 214.

10. Daniel Tiffany, *Toy Medium: Materialism and Modern Lyric* (Berkeley: University of California Press, 2000), 226.

11. The invisible or transparent man in *The Invisible Man* has been thus rendered as a result of experiments conducted by the Japanese Imperial Army on its own soldiers. His rage against militarism in general and Japanese militarism in particular signals the move toward an explicit reference to the events of World War II and a departure from the fantastic and allusive nature of the earlier *The Invisible Man Appears.*

12. A number of films made after 1952, such as Mizoguchi Kenji's 1953 *Ugetsu (Ugetsu Monogatari)* and Honda Ishirô's 1954 *Godzilla (Gojira),* are frequently read as allegories of World War II, as attempts to render what otherwise defies description. It is perhaps important to remember that as a rhetorical device, allegorical structures never restore a unity to the work, but rather continue to divide the work from itself, against itself, producing, in the process, a series of phantom trajectories that move away from the manifest content of a work. Thus outside facts or stated intentions (by authors, for example) cannot always verify the presence of an allegorical subtext, ultimately. Allegories are often intimated, unconscious, or invisible, culled, as it were, from an inaccessible referent.

13. It is also worth noting, perhaps, that the gendered subject "man" has changed to the neutral "human being" *(ningen),* although the invisible or transparent beings remain men in the Japanese films. This erasure of the gendered figure in the title is interesting in part because the economies of sexuality and eros continue to play an active role in the Japanese films. In a subplot of the 1949 film, a member of the all-female Takarazuka revue appears in a number of scenes, including a cabaret performance in which she portrays a masculine figure. Later, the Takarazuka star, Mizuki Ryūko, who is also the younger sister of Kurokawa, the invisible man, masquerades as the invisible man—in bandages and sunglasses—in an effort to rescue the kidnapped scientist.

An erotic cabaret performance also appears in the 1954 version at a nightclub named "Kurofune," or "black ship," which was the name given by the Japanese to Commodore Matthew Perry's American ships when they approached the shores of a then closed Japan on 8 July 1853. (One hundred years before *The Invisible Man* on 31 March 1854, the United States and Japan signed a treaty opening some of Japan's ports to American vessels.) The black ships came to be seen as, among other things, symbols of an eroticized foreignness, blending a libidinal interest in and fear of the other. In *The Invisible Man Meets the Fly,* the owner of the nightclub Ajia (Asia) is named Kuroki. This installment of the series also features extensive erotic cabaret performances. Some form of the word *black, kuro,* appears in each of the three invisible man films.

14. Murayama later directed fantastic TV programs, including *Ultraman, Ultra Q,* and *Ultra Seven.*

15. Daniel Paul Schreber, a nineteenth-century presiding judge to the Leipzig Court of Appeals in Germany, experienced a severe form of psychosis for which he was hospitalized. One of his symptoms involved an overwhelming fear of sunlight, which he believed was an attempt by God to impregnate him. He saw the rays as divine semen, noting in his memoirs, "Enlightenment rarely given to mortals has been given to me" (D. P. Schreber, *Memoirs of My Nervous Illness,* ed. and trans. Ida MacAlpine and Richard Hunter [Cambridge, MA: Harvard University Press, 1988], 167). Freud studies the Schreber case in "Psycho-Analytic Notes on an Autobiographical Account of a Case of Paranoia (Dementia Paranoides)," in *The Standard Edition of the Complete Psychological Works of Sigmund Freud,* ed. and trans. James Strachey (London: Hogarth, 1953–74), vol. 12. Of Schreber's photophobia, Lacan suggests that the "rays, which exceed the bounds of recognized human individuality" and which are "unlimited," are capable of annihilating not only individual human beings but human subjectivity itself (Jacques Lacan, *The Seminar of Jacques Lacan: The Psychoses 1955–1956,* ed. Jacques Alain-Miller, trans. Russell Grigg [New York: Norton, 1993], 3:26).

16. Two notions of interiority are at work, physical and psychological. Monocaine appears to unite those two forms of interior human space. It suggests a "single, unitary, and atomic" element, says Albert Liu (letter to author, March 2004).

17. The exchange of visibility for life remains an unexplained trope in most "invisible man" works. Once rendered invisible, the afflicted organism regains visibility only at the moment of death. This synchronicity, repeated in the novel and

film versions, puts into play an economy of visuality and life, a bio-optic circuit that establishes invisibility as a temporary condition, unsustainable, in the end, as a mode of living.

18. Michel Chion, *Audio-Vision: Sound on Screen*, trans. Claudia Gorbman (New York: Columbia University Press, 1994), 128.

19. Wells, *Invisible Man*, 146.

20. Ibid., 146 (original emphasis).

21. Ibid.

22. Liu, letter to author.

23. Wells, *Invisible Man*, 208.

24. Ibid., 207.

25. Ibid. Later in the novel, Wells/Griffin refers to his landlord as "an old Polish Jew in a long grey coat and greasy slippers" (217).

26. Ibid., 207 (emphasis added).

27. Ibid., 208.

28. Ibid., 209–10 (original emphases).

29. Ibid., 211.

30. Ibid.

31. Ibid.

32. Ibid., 213–14.

33. Ibid., 214. An invisible cat also appears in *Tômei ningen arawaru*.

34. Ibid., 215.

35. Ibid., 219.

36. Ibid.

37. Ibid., 236.

38. Ibid., 236 (emphasis added).

39. Ibid.

40. Ibid., 274.

41. Ibid., 274–75.

42. In a 1955 report "Chorioretinitis Produced by Atomic Bomb Explosion," Dr. Jacques Landesberg describes the case of an Army second lieutenant who witnessed an atomic explosion test at Yucca Flats. Despite instructions not to look directly into the explosion, the soldier peered over his shoulder at the blast and saw "a very bright, white, blinding light." For several days afterward, he experienced partial blindness in one eye. On closer examination, doctors discovered a retinal burn. "The shape," writes Landesberg, "is roughly that of an inverted mushroom" (*Archives of Ophthalmology* 54 [October 1955]): 539–40). In an otherwise strictly scientific idiom, Landesberg allows for this figure. The "mushroom," it seems, had burned itself directly onto the soldier's eye.

43. Paul Virilio, *War and Cinema: The Logistics of Perception*, trans. Patrick Camiller (London: Verso, 1989), 81 (original emphasis).

44. The function and properties of writing as a form of staining play a critical role in the dynamic of atomic representation. Freud speaks of psychoanalysis as a kind of staining of the unconscious, a gesture that brings into relief its invisible contours. In a similar manner, one finds "scenes of writing," as Derrida might call

them, in a number of postwar Japanese films. In many such instances, the act of writing brings the invisible form to light, while the act of erasing, which frequently accompanies it, thrusts the body back into the recesses of darkness. In *The Invisible Man,* the transparent person "passes" his days as a circus clown. This form of body writing or painting allows him to participate in the visible world as invisible. When he decides to reveal his true form to a journalist, he wipes the makeup from his face, in effect, erasing himself from sight.

45. Virilio, *War and Cinema,* 81. Of the relation between war and perception, Virilio says, "*The history of battle is primarily the history of radically changing fields of perception....* War consists not so much in scoring territorial, economic, or other material victories as in appropriating the 'immateriality' of perceptual fields" (7, original emphasis).

46. In this sense, the Japanese "fly" resembles more closely the irradiated shrinking man of Jack Arnold's 1957 film *The Incredible Shrinking Man* than it does the insect-human hybrid of Kurt Neumann's 1958 film *The Fly.* During World War II, the U.S. military experimented with "bat bombs," incendiary devices attached to living bats, which were designed to ignite fires in Japan. The plan was to drop bat bombs over Japanese cities and then trigger the bombs once the bats had sought refuge in wooden Japanese homes. After several unsuccessful tests, the project was suspended and ultimately rendered obsolete with the advent of the atomic bomb. The Department of the Navy called the bat bomb experiments "Project X-Ray."

47. Liu, letter to author.

48. In *Audio-Vision,* Michel Chion considers *The Invisible Man* one of the first great sound films. "The impact of *The Invisible Man,*" says Chion, "stems from the cinema's discovery of the powers of the invisible voice.... The speaking body of Wells's hero Griffin is not invisible by virtue of being offscreen or hidden behind a curtain, but apparently really in the image, even—and above all—when we don't see him there" (126, original emphasis). Chion calls such an "invisible voice" acousmatic—phantom sounds whose sources are not visible. They participate in the visible world as invisible, linking audio to vision through displacement. Griffin's "downfall and death," says Chion, "is linked to his common fate of visibility"; the collapse of his avisuality marks his return to the visible world and his death (127).

49. See Shuhei Hosokawa, "Atomic Overtones and Primitive Undertones: Akira Ifukube's Sound Design for *Godzilla,*" in *Off the Planet: Music, Sound, and Science Fiction Cinema,* ed. Philip Hayward (London: Libbey, 2004), 42–60.

50. Wells, *Invisible Man,* 223.

51. The insufficiencies of language are etched into the very appellation of the atomic moment. The authors of *The Day Man Lost: Hiroshima, 6 August 1945* explain: "Those who did not hear the bomb called it *pika*—'the flash'; those who did hear it called it the *pikadon*—'the flash-boom'" (Pacific War Research Society, *The Day Man Lost: Hiroshima, 6 August 1945* [Tokyo: Kodansha International, 1972], 238). The recourse to a mimetic, onomatopoetic language underscores the radically *photographic* effects of the atomic explosions—they have left their imprints on a language that is unable to describe them.

52. In *Atomic Bomb Cinema,* Jerome F. Shapiro notes Honda Ishirô's use in

Godzilla of wounds to the eye and figures of blindness as symbols of war. Of the one-eyed scientist Serizawa, Shapiro writes: "According to Mr. Honda, in my interview with him in 1990, Serizawa's missing eye is—at least to 1950s Japanese audiences—an emblem, or stigma, of his wartime experience and suffering" (273). The wounded eye would have been understood as a signifier of the war.

53. De Kooning, "What Abstract Art Means to Me," 60.

54. Giorgio Agamben, *Means without End: Notes on Politics,* trans. Vincenzo Binetti and Cesare Casarino (Minneapolis: University of Minnesota Press, 2000), 92 (original emphases). "The face is at once the irreparable being-exposed of humans and the very opening in which they hide and stay hidden" (91).

55. Ibid., 91.

56. Ibid., 95.

57. Gilles Deleuze and Félix Guattari, *A Thousand Plateaus: Capitalism and Schizophrenia,* trans. Brian Massumi (Minneapolis: University of Minnesota Press, 1987), 168 (emphasis added). "Or should we say things differently? It is not exactly the face that constitutes the wall of the signifier or the hole of subjectivity. The face, at least the concrete face, vaguely begins to take shape *on* the white wall. It vaguely begins to appear *in* the black hole. In film, the close-up of the face can be said to have two poles: make the face reflect light or, on the contrary, emphasize its shadows to the point of engulfing it 'in pitiless darkness'" (168, original emphases).

58. Agamben, *Means without End,* 95.

59. Ibid., 99–100 (original emphases).

60. As commented by a member of the discussion at the symposium "The Face of Another: Japanese Cinema/Global Images" (Yale University, 2002).

61. For more on the invisibility of Koreans and other minorities in Japan after the war, see Lisa Yoneyama, "Ethnic and Colonial Memories: The Korean Atom Bomb Memorial," in *Hiroshima Traces: Time, Space, and the Dialectics of Memory* (Berkeley: University of California Press, 1999), 151–86. Yoneyama describes the memorial for Korean victims of the atomic bomb, erected in Hiroshima in 1970, "as one of the very few visible reminders of the tribulations and suffering of those minorities interpellated as Koreans" (153).

62. The absence of the definite particle, "the," erased in Ellison's version may refer to another mode of invisibility that divides the subject itself and from itself. Invisibility in Ellison's usage refers to one's visibility, but also the infinite divisibility, perhaps, of the individual. No longer indivisible, which is to say, singular, an invisible man is also no longer individual. He is divisible, atomic. Ellison's invisible man functions as a pseudohomonym for Wells's invisible man. Buried within the letters that constitute the expression "invisible man" is the quasi-anagrammatic sentence "I am."

63. Ellison was a merchant marine on sick leave when the war ended. Ellison's *Invisible Man* begins, in a sense, with the end of World War II, in its wake.

64. Ralph Ellison, *Invisible Man* (New York: Vintage, 1980), xv. Fred Moten says, "The mark of invisibility is a visible, racial mark; invisibility has visibility at its heart. To be invisible is to be seen, instantly and fascinatingly recognized as the unrecognizable, as the abject, as the absence of individual self-consciousness, as a

transparent vessel of meanings wholly independent of any influence of the vessel itself" (Fred Moten, *In the Break: The Aesthetics of the Black Radical Tradition* [Minneapolis: University of Minnesota Press, 2003], 68). Moten invokes a deeply corporeal invisibility that "has visibility at its heart." What is invisible in Ellison is deeply material, racial, and superficial and is "instantly recognized as the unrecognizable." Ellison's figure moves in an avisual economy, where he is figured and perceived as invisible. In his analysis of Ellison's *Invisible Man*, Moten links music, sound, and noise to invisibility, rendering the impossibility of seeing (or of listening) in Ellison to a persistent displacement of sense, what Moten describes in the black avantgarde as an "ensemble of the senses." Visibility returns in the acoustic, which "*is* seeing," Moten says, and makes the abject visibility of the invisible, avisual (67, original emphasis).

65. Ellison, *Invisible Man,* 3.

66. Ibid. In her chapter, "Materializing Invisibility as X-ray Technology: Skin Matters in Ralph Ellison's *Invisible Man,*" Maureen F. Curtain argues against an exclusively metaphoric reading of invisibility in Ellison's novel. Curtain argues for the sustained activity of X-rays in the novel, operating as a trope, figure, and technology that insists on the invisible man's invisibility as a "bio-technical phenomenon." She says: "Invisibility signals not the end of materiality, but rather only the disappearance of the skin" (Maureen F. Curtain, *Out of Touch: Skin Tropes and Identities in Woolf, Ellison, Pynchon, and Acker* [New York: Routledge, 2003], 43).

67. Ellison, *Invisible Man,* 3.

68. Moten, *In the Break,* 84. Again, in Moten's idiom, the invocation of the heart, a figure that opens onto a corporeal and phantasmatic geography.

69. Ellison, *Invisible Man,* 3.

70. Ibid., 6 (original emphasis).

71. Ibid., 6 (original emphasis).

72. Ibid., 7. The invisible man continues: "To be unaware of one's form is to live a death. I myself, after existing some twenty years, did not become alive until I discovered my invisibility" (7).

73. Ellison, *Invisible Man,* 438. After witnessing Clifton's murder at the hands of a New York City policeman, the invisible man asks: "Why did he choose to plunge into nothingness, into the void of faceless faces, of soundless voices, lying outside history?" (439). The invisible man comes to understand Clifton's plunge outside history.

74. Ibid., 474 (emphasis added).

75. Ibid., 474.

76. Moten, *In the Break,* 201.

77. Ibid.

78. Deleuze, *Logic of Sense,* 156.

79. Moten, *In the Break,* 200.

80. Deleuze, *Logic of Sense,* 157.

81. Moten's *In the Break* is an extended treatise on the avisuality of "blackness," a phenomenon exposed in the collapse of the audiovisual divide or screen, which Moten locates in "the Black Radical Tradition." In his moving analysis of the

disfigured face of Emmett Till, revealed at his open casket funeral and captured in the photographs that documented it, Moten argues for the impossibility of separating photography from phonography. "Emmett Till's face is seen," Moten says, "was shown, shone. His face was destroyed (by way of, among other things, its being shown: the memory of his face is thwarted, made a distant before-as-after effect of its destruction, what we would have otherwise seen). It was turned inside out, ruptured, exploded, but deeper than that it was opened" (198). The opening of Till's face, the injustice and violence it exposes, unleashes a sound, a cry, an echo of the "whistle" that sealed his fate, the irrepressible and visible sound that Moten calls "black mo'nin'." He says: "Looking at Emmett Till is arrested by overtonal reverberations; looking demurs when looking opens onto an unheard sound that the picture cannot secure but discovers and onto all of what might be said to mean that I can look at his face, this photograph" (198). The rupture of Till's face, "turned inside out," turns the world inside out: image and sound, photography and phonography, this world and its other, utopia, are rendered visible. Everything is made visible, especially the sound of black mo'nin'. It is the acousmatic visuality of this photograph.

82. Ellison, *Invisible Man,* 239–40.

83. Deleuze, *Logic of Sense,* 10.

84. Dragan Kujundzic and David Theo Goldberg organized a symposium titled "tRACEs: Race, Deconstruction, and Critical Theory" at the University of California, Irvine, Humanities Research Institute in April 2003. On the relationship between race and all of the erasures that constitute its histories, Kujundzic and Goldberg say, in their announcement: "These traces conjure memory in advance: at once urgent and untimely, they expose themselves and take a chance with time. If *apartheid,* as Derrida pointed out nearly twenty years ago, will come to be the name of something finally abolished, the site of a history faded in memory, much as antisemitism came to be for critical theory, then racelessness comes to be the future of a present whose racial traces are at once silenced and silently mark social life worlds, the rearview vision of the future. But, Derrida reminded us never to forget, 'hasn't apartheid always been the archival record of the unnameable?' Confined and abandoned to the silence of this memory, the name will resonate all by itself, reduced to the state of a term in disuse. The thing it names today will no longer be, but the tomorrow it suggests calls for critical engagement now. 'tRACEs' accordingly is a wager about the future of the present."

85. Partially blind cult leader Asahara Shōko predicted that the world would be destroyed by nuclear weapons in 1999.

86. Shinoda's narrator compares images of destruction at Kobe in 1995 and throughout Japan in 1945 in *Moonlight Serenade* (*Setouchi mûnraito serenâde,* 1997).

5. *Excription*/Antigraphy

1. Stan Brakhage's 1971 film *The Act of Seeing with One's Own Eyes* chronicles the activities of a Pittsburgh morgue, following in graphic detail the various aspects and stages of multiple autopsies. This film establishes at the locus of the

screen the brutal divides between one's eyes and another's, the camera view and the unlooking glare of radical exteriority.

2. Maurice Merleau-Ponty, "Eye and Mind," in *The Primacy of Perception: And Other Essays on Phenomenological Psychology, the Philosophy of Art, History, and Politics,* ed. James M. Edie, trans. Carleton Dallery (Evanston, IL: Northwestern University Press, 1964), 162. "It sees itself seeing; it touches itself touching; it is visible and sensitive for itself. It is not a self through transparence, like thought, which only thinks its object by assimilating it, by constituting it, by transforming it into thought" (162–63).

3. Ibid., 163.

4. Ibid., 164 (emphasis added).

5. Ibid., 164.

6. Ibid.

7. Ibid.

8. Ibid., 166. In *Francis Bacon,* Gilles Deleuze makes a similar claim. Following Paul Klee, who said of painting, "Not to render the visible, but to render visible," Deleuze says: "The task of painting is defined as the attempt to render visible forces that are not themselves visible. Likewise, music attempts to render sonorous forces that are not themselves sonorous (Gilles Deleuze, *Francis Bacon: The Logic of Sensation,* trans. Daniel W. Smith [Minneapolis: University of Minnesota Press, 2003], 48).

9. Merleau-Ponty, "Eye and Mind," 169.

10. Jun'ichirô Tanizaki, *In Praise of Shadows,* trans. Thomas J. Harper and Edward G. Seidensticker (New Haven, CT: Leete's Island Books, 1977), 31–32.

11. Ibid., 31.

12. Ibid., 32.

13. Ibid.

14. Ibid. "The sight offends even our own eyes," he concludes, "and leaves none too pleasant a feeling."

15. Merleau-Ponty, "Eye and Mind," 168.

16. Samuel Weber, "Mass Mediauras, or: Art, Aura, and Media in the Work of Walter Benjamin, in *Mass Mediauras: Form, Technics, Media,* ed. Alan Cholodenko (Stanford, CA: Stanford University Press, 1996), 88. Weber finds in Heidegger's and Benjamin's recourse to the language of images, of the picture *(Bild),* two theses on the shadow that resemble, in Weber's analysis, that of Tanizaki. Of Heidegger's essay "The Age of the World Picture," Weber says: "Once the human has been determined as subject and the world as picture, Heidegger remarks, an 'invisible shadow is cast over all things,' a shadow which prevents them from ever being put fully into their proper places, that is, being fully depicted. This shadow is not simply external to the world as picture; it is an inseparable part of it. The world as picture reveals itself—which is to say conceals itself—as shadow. But *shadow* here does not name 'simply the lack of light,' or even less 'its negation.' It designates that which escapes and eludes the calculating plans of total representation, of which it at the same time is the condition of possibility: 'In truth the shadow bears overt and yet impenetrable witness to the concealed glow" ("Mass Mediauras," 81, original emphasis).

Weber's reading of the residual metaphysics of Heidegger's shadow produces a form of avisuality: the shadow evades "total representation," but remains an element of its possibility; the "concealed glow" emerges into the world as a "bright shadow."

17. Martin Heidegger, "The Age of the World Picture," in *The Question concerning Technology and Other Essays,* trans. William Lovitt (New York: Harper and Row, 1977), 154. Writing in 1938, several years after Tanizaki, Heidegger extends his reflections on the shadow to the "incalculable" nature of Being, which "withdraws" from representation. "In keeping with this concept of shadow, we experience the incalculable as that which, withdrawn from representation, is nevertheless manifest in whatever is, pointing to Being, which remains concealed" (154).

18. Tanizaki, *In Praise of Shadows,* 35.

19. Ibid., 32.

20. Ibid., 30. Tanizaki's discourse is here ambiguous, perhaps ambivalent, rehearsing a common rhetorical gesture of the period that seeks to align Japanese sensibilities with Eastern (most often Chinese) aesthetics. See, for example, Okakura Kakuzo's 1906 *The Book of Tea.*

21. Tanizaki, *In Praise of Shadows,* 33.

22. Jacques Derrida, "NO APOCALYPSE, NOT NOW (full speed ahead, seven missiles, seven missives)," trans. Catherine Porter and Philip Lewis, *Diacritics* 14.2 (1984): 20–31.

23. John Treat says the atomic bombings of Hiroshima and Nagasaki produced *memories* and *images of memories* of the end of the world: "Since the destruction of the two cities through the use of nuclear weapons in August 1945, some of us have the memory—and the rest of us, our imagination of that memory—of how the world may end" (John Whittier Treat, *Writing Ground Zero: Japanese Literature and the Atomic Bomb* [Chicago: University of Chicago Press, 1995], 1). Two imaging devices, memory and imagination, two modes of avisuality, summoned in the face of total invisibility.

24. Maurice Blanchot, *The Writing of the Disaster,* trans. Ann Smock (Lincoln: University of Nebraska Press, 1986), 41.

25. Ibid., 1.

26. Masuji Ibuse, *Black Rain,* trans. John Bester (Tokyo: Kodansha International, 1969), 34.

27. Ibid., 35.

28. Treat, *Writing Ground Zero,* 8.

29. At work in Tanizaki's discourse is the desire for a kind of alternative physics of the body. He writes: "I always think how different everything would be if we in the Orient had developed our own science. . . . The Orient quite conceivably could have opened up a world of technology entirely its own" (7).

30. George Weller, an American journalist who slipped undetected into Nagasaki several days after the atomic bombing of the city in 1945, wrote a series of stories that described the effects of the bombing and the radiation sickness it caused among survivors. Weller's reports were ultimately rejected by U.S. military censors, lost, and rediscovered sixty years later. In an idiom reminiscent of Röntgen's, Weller described the radiation sickness as "disease X."

31. Jean-François Lyotard has described a similar rhetorical impasse, which he calls a "differend." For Lyotard, the differend arises from a conflict in which no single rule can accommodate two or more competing discourses. It suggests that "a universal rule of judgment between heterogeneous genres is lacking in general" (Jean-François Lyotard, *The Differend: Phrases in Dispute,* trans. Georges Van Den Abbeele [Minneapolis: University of Minnesota Press, 1988], xi). The attempt to reconcile the atomic bombings with any known discourses has produced something like a differend.

32. Jean-Claude Lebensztejn suggested this line of thinking concerning emulsions.

33. Vivian Sobchack, *The Address of the Eye: A Phenomenology of Film Experience* (Princeton, NJ: Princeton University Press, 1992), 59 (original emphasis).

34. Ibid., 59.

35. To this she responds: "I saw everything." Total visibility and invisibility, everything and nothing are at stake in the visuality of Hiroshima.

36. The history of tattooing in Japan undoubtedly informs the trope of body writing in some way. The influence of tattooing on popular Japanese culture and what is generally understood from the outside as a Japanese obsession with surfaces (i.e., Roland Barthes), creates a context for the atomic trope. See Donald McCullum's analysis of tattoos and Japanese culture, "Historical and Cultural Dimensions of the Tattoo in Japan," in *Marks of Civilization: Artistic Transformations of the Human Body,"* ed. Arnold Rubin (Los Angeles: Museum of Cultural History, University of California, Los Angeles, 1988), 109–34.

37. Siegfried Kracauer, *Theory of Film: The Redemption of Physical Reality* (New York: Oxford University Press, 1960), 71.

38. Albert Liu, letter to author, March 2004.

39. Dudley Andrew reads the political and phenomenal play of invisibility in another of Mizoguchi's films from the post-Occupation period (1945–52), *Sanshô the Bailiff* (*Sanshô Dayû,* 1954): "If one takes Japan as a nation that during the Occupation became invisible to itself—when the United States prevented the past from being seen and revoked customs and traditions—then *Sanshô Dayû* could be said to call this people to consciousness" (Dudley Andrew, "Mizo Dayû," in *Sanshô Dayû,* Dudley Andrew and Carole Cavanaugh [London: British Film Institute, 2000], 44).

40. Giorgio Agamben, following Primo Levi, describes the state of certain prisoners of the death camps who were labeled "Muselmänner," or Muslims. They were the living dead of the camps. Many who reached this condition—listless and aimless, virtually without language: a "faceless presence," says Levi—died (Primo Levi, *Surviving in Auschwitz and The Reawakening: Two Memoirs,* trans. Stuart Woolf [New York: Summit Books, 1986], 90, cited in Giorgio Agamben, *Remnants of Auschwitz: The Witness and the Archive,* trans. Daniel Heller-Roazen [New York: Zone, 1999]). But some returned. Those who did returned from the other side of language. "This implies," says Agamben of the ability to survive radical erasure in life, "that in human beings, life bears with it a caesura that can transform all life into survival and all survival into life" (*Remnants of Auschwitz,* 133). Life as survival, survival of life: the reversibility of the phrase renders life an emulsion, a mix-

ture of life and its antithesis, all living a form of overcoming life, of always return-ing to life as a condition of life.

41. Fluid imagery saturates the narrative. Although the scene of inscription in Mizoguchi's *Ugetsu* appears in another form in Akinari's original text, both ver-sions make numerous references to liquids and fluidity. In the text, the priest him-self dispels the demon and her attendant: "A spout of water ascended, as if to the sky," Akinari writes, "and the two women disappeared. Clouds engulfed the party and as though spilling black India ink brought down a torrential deluge of rain" (Ueda Akinari, "The Lust of the White Serpent," in *Ugetsu Monogatari: Tales of Moonlight and Rain,* trans. Leon Zolbrod [Tokyo: Tuttle, 1974], 177). Here, the tor-rent replaces the text on the body, descending on Genjuro (Toyoo in the original) and flushing his skin with India ink. While transforming the black rain into callig-raphy, Mizoguchi has retained its aquatic force.

Liquid imagery continues in other scenes throughout Akinari's text. A second priest who tries to exorcise the demons by mixing sulfur with medicine water meets his match when "the creature [opens] its mouth more than three feet wide; its crim-son tongue [darting], as if to swallow the priest in a single gulp" (180). During the encounter, the priest is engulfed by "poisonous vapours" and ultimately lapses into unconsciousness. "His face and body were mottled red and black," writes Akinari, "as though they were stained with dye" (180). The demons, which have metamor-phosed into snakes, are ultimately vanquished, in Akinari's text, by a monk's robe "saturated in mustard-seed incense." In *Ugetsu,* the sensorium that ranges from touch to smell travels through a fluidic economy.

42. Christian Metz, *The Imaginary Signifier: Psychoanalysis and Cinema,* trans. Celia Britton, Annwyl Williams, Ben Brewster, and Alfred Guzzetti (Bloomington: Indiana University Press, 1982), 101. Of the relation between dreams and films, Metz says: "We sometimes speak of the illusion of reality in one or the other, but true illu-sion belongs to the dream and to it alone. In the case of the cinema, it is better to limit oneself to remarking the existence of a certain *impression* of reality" (101).

43. "The dream proper is an image based on the movement of sense impres-sions" (Aristotle, "On Dreams," in *The Complete Works of Aristotle,* ed. Jonathan Barnes, trans. J. I. Beare [Princeton, NJ: Princeton University Press, 1984], 1:735).

44. Ueno Ichirô, "Review," in *Ugetsu: Kenji Mizoguchi, Director,* ed. Keiko I. McDonald (New Brunswick, NJ: Rutgers University Press, 1993), 118.

45. Lafcadio Hearn, "The Story of Mimi-Nashi-Hôichi," in *Kwaidan: Stories and Studies of Strange Things* (Tokyo: Tuttle, 1971), 3.

46. Jacques Derrida, *Memoirs of the Blind: The Self-Portrait and Other Ruins,* trans. Pascal-Anne Brault and Michael Naas (Chicago: University of Chicago Press, 1993), 126 (original emphases).

47. Ibid., 127.

48. Gilles Deleuze, *Cinema 1: The Movement-Image,* trans. Hugh Tomlinson and Barbara Habberjam (Minneapolis: University of Minnesota Press, 1986), 79. "Water," says Deleuze, "is the most perfect environment in which movement can be extracted from the thing moved, or mobility from movement itself" (77).

49. Maureen Turim suggested this comparison.

50. David Serlin describes the visit of the Hiroshima Maidens: "Unlike survivors of conventional war, the Maidens elicited an unprecedented outpouring of medical anxieties and fears concerning the treatment of the body (and, especially, the female body) as disabled by the effects of radiation and nuclear fallout. Much to the State Department's chagrin, the visual and symbolic evidence of the Maidens' damaged bodies in public view helped to forge unavoidable links between their physical scarring and the damage wrought by the atomic bomb. In newspapers across the country, the young women were described variously as 'bomb-scarred,' 'A-scarred,' 'Hiroshima-scarred,' 'A-burned,' 'Atomic-bomb-scarred,' or simply as 'A-girls' or 'A-victims'" (David Serlin, "The Clean Room/Domesticating the 'Hiroshima Maidens,'" *Cabinet* 11 [summer 2003]: 8).

51. Peter Greenaway's 1996 film *The Pillow Book,* a loose adaptation of Sei Shônagon's book of the same title, makes extensive use of the trope of body writing to play ironically with the Japaneseness of a film that is only nominally Japanese.

52. The assemblage, they say, a *disorganized* anti-body, "works only through the dismantling [*démontage*] that it brings about on the machine and on representation. And, actually functioning, it functions only through and because of its own dismantling. It is born from this dismantling" (Gilles Deleuze and Félix Guattari, *Kafka: Toward a Minor Literature,* trans. Dana Polan [Minneapolis: University of Minnesota Press, 1986], 48).

53. Jean-François Lyotard invents this term, "acinema," to describe the general practice of "effacement and exclusion" in filmmaking that leads inevitably to abstraction (Jean-François Lyotard, "Acinema," in *Narrative, Apparatus, Ideology: A Film Theory Reader,* ed. Philip Rosen, trans. Paisley N. Livingston [New York: Columbia University Press, 1986], 349–59).

54. Blanchot, *Writing of the Disaster,* 1.

55. Tanizaki, *In Praise of Shadows,* 30.

56. Martin Arnold erases, in *Deanimated: The Invisible Ghost* (2002), characters from a horror film, Joseph H. Lewis's 1941 *The Invisible Ghost.* After digitizing the film, Arnold selectively and sequentially removes characters from the original, leaving behind empty props, uninhabited sets, and seemingly aimless camera movements. The backgrounds are replaced, so the erasure leaves no trace. The effects are not unlike some aspects of invisible man films except that Arnold also erases in some places the voices of characters. In these instances, Arnold seals the characters' mouths shut, leaving them with unexplained bodily tremors and uncomfortably long nonresponses. Arnold's vocal erasures inscribe the trope of erasure, an avisuality of the voice, signaled by a silent body that trembles. In her recycled film *Removed* (1999), Naomi Uman also erases characters from found footage pornography. She has stained the work in a pink nail polish, then erased the figures of women with a remover, leaving behind tactile white shadows where the women once were.

57. Blanchot, *Writing of the Disaster,* 7.

58. Abe Kôbo, *The Woman in the Dunes,* trans. E. Dale Saunders (New York: Knopf, 1964), 7.

59. Ibid. 3.

60. "Otagai wo tashikameau tame no arayuru shômeisho" (in order to convince one another of our existence).

61. "Otoko mo onna mo aite ga wazato te wo nuiteirunodewa naika to kurai saigi no toriko to naru. Keppaku wo shimesu tame ni muri wo shite atarashî shobun wo omoi tsuku" (men and women become prisoners of their dark suspicions toward one another. To plead their innocence, they think up new documents).

62. In the magazine, the man sees a cartoon and breaks into a convulsive laughter, bordering on hysteria. In the single-frame cartoon, a man, apparently having failed in his diet, has been run over by a steamroller, which has flattened his body below his neck. The image invokes an early film comedy, James Williamson's 1905 *An Interesting Story.*

63. Deleuze and Guattari use the terms "smooth" and "striated" to describe the play of inscription and erasure in space. Their geographic figure for the convergence of writing and erasing is the desert. Smooth and striated spaces always exist as a mixture, as an emulsion: "The two spaces in fact exist only in mixture: smooth space is constantly being translated, transversed into striated space; striated space is constantly being reversed, returned to a smooth space. In the first case, *one organizes even in the desert; in the second, the desert gains and grows; and the two can happen simultaneously*" (Gilles Deleuze and Félix Guattari, *A Thousand Plateaus: Capitalism and Schizophrenia,* trans. Brian Massumi [Minneapolis: University of Minnesota Press, 1987], 474–75, emphasis added).

64. Ibid., 497–98 (original emphasis).

6. Phantom Cures

1. *Maborosi*'s obscure visuality evokes another dark film, David Lynch's 1977 *Eraserhead,* which is similarly composed of dark surfaces. In contrast to *Maborosi,* the strategic luminosity of *Eraserhead* renders the film avisual but not invisible. Its avisuality is generated, it seems, from the acousmatic sound that envelops, permeates, erupts from, and ultimately constitutes the film's visual field. Besides the figure of erasure named in its title and embodied in its protagonist, *Eraserhead* also bears the inscription "x": Lynch's female lead, the mother of Henry's child, is named in the credits as Mary X, her parents Mr. and Mrs. X.

2. Kore-eda, in an interview with Gabriel M. Paletz, describes his reluctance to use flashbacks, linking it to his background in documentary filmmaking. About his first uses of flashbacks in his third feature film *Distance* (2001), Kore-eda says: "I've always avoided using flashbacks, as memories presented in real images. . . . The reason for avoiding concrete images from the past is that to make a documentary, my basic stance is not to express what's inside a subject. Documentaries should try to reveal the inside by showing the outside" (Gabriel M. Paletz, "The Halfway House of Memory: An Interview with Hirokazu Kore-eda," *CineAction* 60 [2003]: 58).

3. In his discussion with Paletz, Kore-eda describes the role of the green bicycle as an extension of the film's entire color: "*Maborosi*'s color was quite carefully planned. In the first half of the film, the main tone is green. So Yumiko and Ikuo even paint the bicycle green" (Paletz, "Halfway House of Memory," 57). Green objects set against the green tones of the film also flatten the world onto the screen.

4. In this sense, Mamiya is not unlike the vacant man in Alain Resnais's 1962 film *Last Year at Marienbad* named "X," who seeks to convince a woman "A" that they had met one year earlier. He arrives to her like an externalized memory, no trace of which remains within her. "You must remember," he presses her, as he recounts to her the episodes of their earlier encounter. He is at once an X-ray and erasure. "You are like a shadow," she says to him. Emptiness as a figure for invisibility also appears in Paul Verhoeven's 2000 film *Hollow Man,* which employs the idiom of hollowness as a synonym for invisibility, suggesting a relationship between volume and visibility.

5. Kurosawa Kiyoshi, *Eiga wa osoroshî* (Tokyo: Seidôsha, 2001), 302. Kurosawa describes his method. He selected large and empty rooms for his shoots, then filled them with simple and ordinary furniture and asked his actors to act in an ordinary and everyday manner. The strangeness of the scenes emerges from the great distances, as much as ten meters, between the pieces of furniture and the elongated and emptied spaces generated by the actors who moved between them (302).

6. Sakuma's phrase "don't take me too seriously" *(ma ni ukeruna)* echoes his earlier explanation of the crimes to Takabe, "the devil made them do it" *(ma ga sashita).* "Ma" can mean either "truth" or "evil." Sakuma's name rhymes with *akuma,* or evil; his first name, Makoto, means "truth." The word or character "ma" in both Mamiya and Sakuma has yet another meaning, "interval" or "space," a term used extensively in the Noh theater.

7. Kurosawa, *Eiga wa osoroshî,* 301.

8. Mamiya's gesture echoes the movements of the hand signing "x" in the historical footage that Sakuma shows Takabe. A technical gesture (for inducing hypnosis) and a signature (signing with an imaginary stylus), the "x" sign also represents the act of cutting another open, of entering and exiting another. In *Of Grammatology,* Jacques Derrida locates a gesture of signing, what he calls "mute signing," in Rousseau. Rousseau refers to the gesture as a "magic wand." Derrida says: "The movement of the magic wand that traces with so much pleasure does not fall outside the body. Unlike the spoken or written sign, it does not cut itself off from the desiring body of the person who traces or from the immediately perceived image of the other. . . . She who traces, holding, handling, now, the wand, is very close to touching what is very close to being the other *itself,* close by a minute difference; that small difference—visibility, spacing, death—is undoubtedly the origin of the sign and the breaking of immediacy (Jacques Derrida, *Of Grammatology,* trans. Gayatri Chakravorty Spivak [Baltimore, MD: Johns Hopkins University Press, 1976], 234, original emphasis). The "magic wand" or "mute sign," unlike the "spoken or written sign," does not leave the body entirely. It forms a virtual contact between two bodies, separated "by a minute difference" and serves as the "origin of the sign." In *Cure,* this mute sign connects two bodies, the movement of one into another, without a complete transfer of subjectivity, one to the other. The desiring body remains inside the other, a secret transaction or touch, a shadow transference.

9. Sigmund Freud, "Moses and Monotheism: Three Essays," in *The Standard Edition of the Complete Psychological Works of Sigmund Freud,* ed. and trans. James Strachey (London: Hogarth, 1955), 23:63.

10. The trope of transmission from Freud to *Cure,* hypnosis to missionary conversion, also establishes the force of another Japanese horror film, Nakata Hideo's 1998 *Ring,* which is based on the imperative to transmit. A *transmissionary* zeal, one could say.

11. "Writing, the trace, inscription, on an exterior substrate or on the so-called body proper, as for example, and this is not just any example for me, that singular and immemorial archive called *circumcision,* and which, though never leaving you, nonetheless has come about, and is no less exterior, *exterior right on* your body proper" (Jacques Derrida, *Archive Fever: A Freudian Impression,* trans. Eric Prenowitz [Chicago: University of Chicago Press, 1996], 26, original emphases).

12. Freud, "Moses and Monotheism," 112–13.

13. Nicolas Abraham and Maria Torok, "Mourning *or* Melancholia: Introjection *versus* Incorporation," in *The Shell and the Kernel: Renewals of Psychoanalysis,* ed. and trans. Nicholas T. Rand (Chicago: University of Chicago Press, 1994), 126 (original emphasis).

14. Ibid., 126.

15. Ibid., 127.

16. "Ana-lysis meaning," says Albert Liu, "breaking down, atomizing, dissolving" (letter to author, March 2004).

17. Paul Virilio, *The Vision Machine,* trans. Julie Rose (Bloomington: Indiana University Press, 1994), 59.

18. Ibid., 75.

19. Ibid., 72–73 (original emphases). Virilio continues: "Seeing and non-seeing have always enjoyed a relationship of reciprocity, light and dark combining in the *passive* optics of the camera lens" (73, original emphasis).

20. Raymond Bellour describes the passage between the world and images as analogy; the relationship is analogic, determined by the logic of natural resemblance. "If we use the concept of nature, in keeping with its religious origin, to designate the relationship of dependence between two terms, the world and the image, the 'analogy' also leads us to presuppose such a relationship between the images themselves, in other words, between forms of images as well as between forms and world(s)" (Raymond Bellour, "The Double Helix," in *Electronic Culture: Technology and Visual Representation,* ed. Timothy Druckrey [New York: Aperture, 1996], 177). For Bellour, the relationship between the world and its images, founded on concepts of religious and natural order, evolves in time to a new mode of analogy, which establishes, like Virilio's vision machines, relationships between images, passages between forms of images and modes of visuality.

21. Virilio calls this moment of the end of visuality, a moment that has just begun, an "intensive eternity." It is a form of atomic temporality that follows from the atomic destruction of matter in space. "After the nuclear disintegration of *the space of matter,*" he says, "which led to the implementation of a global deterrence strategy, the disintegration of *the time of light* is finally upon us" (*Vision Machine,* 72, original emphases). The time of light is atomic, singular and totally destructive, avisual and intensively eternal, a "temporal atom" (72).

Index

Abe Kôbo, 121, 129, 179n.8
Abraham: gift of Isaac to God from, 164n.63
Abraham, Nicolas, 10, 154, 160n.16, 176n.37, 193n.13
absolute density: concept of, 83, 85
absolute invisibility, 32
absorption, 147; Fried's categories of, 65–66, 174n.22
abyss *(mise-en-abîme)*, 56, 69, 175n.27; of encounters, 68–69; of formlessness, 97, 99; in *Maborosi,* 137; of oral cavities, 71, 73; in *Woman in the Dunes,* 129
Achilles, 117
"acinema" based on démontage, 119, 190n.53
acousmatic sound, 135, 138, 142, 182n.48, 191n.1
Acres, Brit, 172n.70
Act of Seeing with One's Own Eyes, The, 185n.1
Adachi Shinsei, 83
Adorno, Theodor, 42, 43, 167n.22, 168n.27

Agamben, Giorgio, 96–97, 161n.25, 162n.22, 183n.54, 188n.40; definition of archive, 11–12
"Age of the World Picture, The" (Heidegger), 186n.16
Agua Radium, 168n.36
AIDS pandemic, 170n.48
albinism of Wells's invisible man, 88, 90, 91–92
Alice's Adventures in Wonderland (Carroll), 177n.38
allegory(ies): of atomic annihilation, 113–20; of avisuality, transformation of photography by X-rays into, 93; in postwar Japanese cinema, 86, 179n.12
allopsy, 105
Âme humaine, ses mouvements, ses lumières et l'iconographie de l'invisible, L' (Baraduc), 172n.80
Amenophis IV, 162n.23
analogy: passage between world and images as, 193n.20
anarchive, 11
anatomical representation, 47–48

195

Index

Weber, Samuel, 107, 186n.16
Weller, George, 187n.30
Wells, H. G., 47–48, 82, 83, 84, 87–92, 98, 99, 143, 178n.3, 179n.9, 183n.62
Whale, James, 87, 178n.3
Williamson, James, 67, 71–73, 175n.27, 176n.37, 191n.62
woman: images of interiority of, 44–47, 172n.80; in Tanizaki's *In Praise of Shadows,* 22–23
Woman in the Dunes (Suna no onna), 121–31; emulsion in, 129, 130, 191n.63; high-contrast cinematography and conflict of dark and light in, 123–24; immiscible mixture of water and sand in, 121, 123–24, 127–28, 129; *mise-en-abîme* in, 129; missing person's report in, 129–31
"Work of Art in the Age of Mechanical Reproduction, The" (Benjamin), 58
World War II: crisis in constitution of human body initiated at end of, 4–5; invisibility of Koreans in Japan after, 98, 183n.60. *See also* Hiroshima and Nagasaki: atomic bombing of
writing: *arkhê*writing, 59; atomic, 27–28, 109–10, 112–13, 121; on body, 1–5, 112–13, 117, 119, 188n.36; *exscriptive,* 55; as form of staining, 181n.44; nondialectical, 113–20; Tanizaki's theory of, 25; transparency as form of representational erasure, 120–21

X-ray photography, 5, 44, 171n.67; X-rays as form of radical photography, 29–30
X-rays, 5, 29–32; atomic radiation and, 83–84; collapse of Enlightenment figure forced by, 42–43; at crossroads of twentieth-century arts and sciences, 168n.29; destructive nature of, 50, 93; discovery of, 42, 44, 46–47, 52, 55; as excess image of Enlightenment sensibility, 80; invisible rendered visible by, 32; as mode of avisuality, 41–53, 55–59; as "New Photography" in 1896, 171n.67; physics and X-ray technology, 50, 170n.51; point of view established by, 42–43; possibilities for organization of interiority offered by, 57–59; prefigured in Freud's "Irma dream," 41, 42; relation between radium and, 171n.58; response to, 52–53, 57, 168n.33; surface of, 80; as technology for visualizing the inside, 30; thereness of, determining, 52–53; trajectory of photography expanded by, 55; transformation of photography into allegory of avisuality, 93; use in art, 53; Visible Human Project (VHP) and, 47–48; as vision machine, 156
X-ray vision, 145–47
"x" sign/signifier, 53–54; in *Cure,* 144, 147, 151–54, 192n.8; as mark of erasure, 147
X: The Man with the X-Ray Eyes, 145–47

Yahweh: murder of Moses and, 15–16
Yerushalmi, Yosef Hayim, 163n.59
Yomota Inuhiko, 98
Yoneyama, Lisa, 183n.60
Yoshimoto, Mitsuhiro, 84, 179n.7

Zehnder, Ludwig, 51

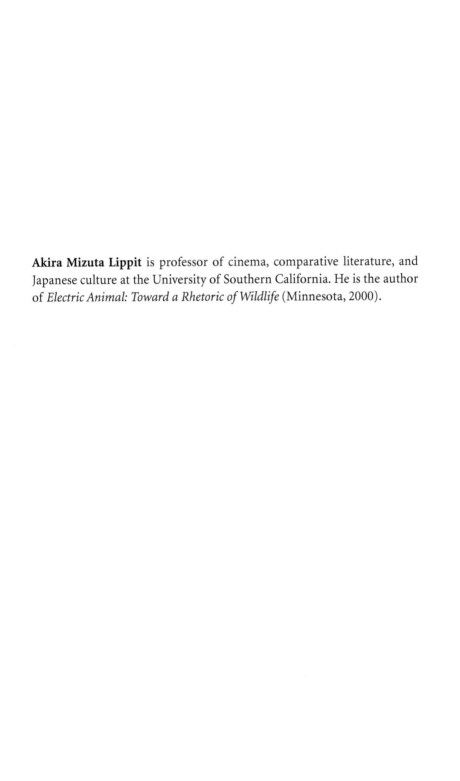

Akira Mizuta Lippit is professor of cinema, comparative literature, and Japanese culture at the University of Southern California. He is the author of *Electric Animal: Toward a Rhetoric of Wildlife* (Minnesota, 2000).